BIOL 110
BIOLOGY I
MOLECULAR AND CELLS

LECTURE GUIDE
AND
LABORATORY MANUAL

5TH EDITION

Karen J. Dalton
Community College of Baltimore County
Biology Department
Catonsville Campus

BIOL 110: Biology I: Molecular & Cells – Lecture Guide & Laboratory Manual, Fifth Edition

Permissions Department
Academx Publishing Services, Inc.
P.O. Box 208
Sagamore Beach, MA 02562
http://www.academx.com

Printed in the United States of America

ISBN-13: 978-1-60036-751-9
ISBN-10: 1-60036-751-8

Table of Contents

PART I: LECTURE & STUDY GUIDE

PART 2: LAB MANUAL

PART I:

LECTURE & STUDY GUIDE

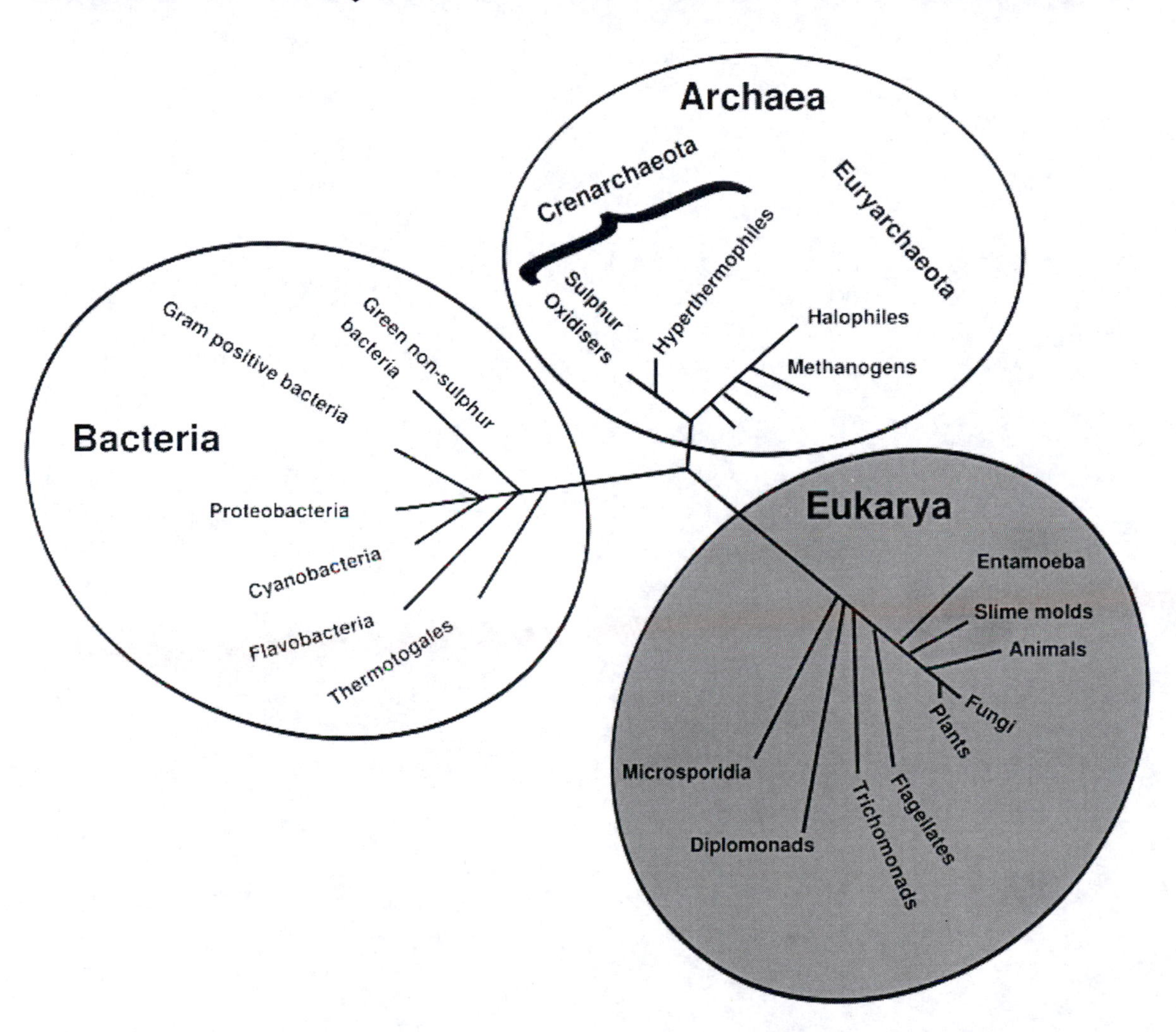

STUDY SKILLS

This is a college course and it will require hard, time consuming work and it will take time. In a 14-week session you will be in lecture for 3 hours a week; therefore, you should be putting in a minimum of 10 hours of study time outside of the classroom. Be prepared to work; but to also have a positive attitude—I can do this.

I. Educational Tools

A. This Course Outline and Workbook

1. To provide the student with an outline of the course. The outline gives the appropriate pages in the textbook(s) where the material can be found and what will be covered on each of the exams.
2. Specific objectives for each unit follow the general outline. Included will be general concepts that should be understood as well as important vocabulary words.
3. Print your instructors Power Point slides—it will also help you in visualizing the material. Add your own notes to these pages, it will complete the picture. There are also usual charts to fill out that will help organize your lecture notes.
4. At home exercises, with answers, have been included. Not all objectives have problems; but the material must still be learned. Using the course outline with the objectives in conjunction with any workbook exercises will help insure success.

B. Textbook

1. Before coming to class, you should have scanned the material that will be covered that day in lecture. Pay particular attention to the headings and boldfaced words. Read the captions under the pictures. Doing this will give you a "jump start" because you with be familiar will new vocabulary words. Review the Power Point slides.
2. Many of the diagrams come directly from the textbook. Again use the diagrams to add your own notes so that the "picture will be complete" in your mind.

C. Study Guide

The study guide contains summaries of the textbook chapters and sample quizzes. Use these as a final review for an exam.

D. Textbook web site (Mind Tap) is an excellent tool to review animations of complex processes. The web site also contains quizzes and other review exercises.

E. Visuals in Library can be found on the library's Reserve Desk.

1. Tapes: either the entire tape (20 to 30 minutes) or small segments of a tape (about 5 minutes)
2. Model Building Kits: hands on activities
3. Additional Computer Programs: interactive programs (review material) some of which contain chapter quizzes.

II. Meet Your Professor

Getting to know your professors is important. Giving your professor a chance to get to know you is equally important. Your goal is to be able to comfortably communicate your concerns, questions, and delights about the course. You are not trying to become teacher's pet nor is your goal to be best friends. This is a professional relationship. You want your professor to have a positive impression of you and to associate your face and your contributions to the course with your name. There are a few simple things that you can do to get know your professor.

1. Sit near the front of the room. This tells your professor that you are interested in the course and are there to learn. Sitting in front makes it easier for the professor to see you. It also increases your probability of being called upon when you raise your hand.

2. Pay attention and participate in class. Professors appreciate and remember the students who participate in their classes. They also remember students who sleep, read the paper, or talk in class. You do not want to be remembered as a disruption.

3. Whenever you speak with your professor, introduce yourself by name (and course if it is outside class time).

4. Go to office hours and introduce yourself early in the semester. Ask if there are any special tips for studying and succeeding in the course. This tells your professor that you are conscientious about your work.

5. Go to office hours throughout the semester. Be sure to take specific questions about the material or assignments. "I was reading in the book and wondered about..." or "In class you said... Could you clarify this for me?". Questions such as "What is going to be on the exam", or "Tell me everything I missed while on my skiing trip" do not get satisfactory answers and waste everyone's time. These sorts of questions give the professor the impression that you really do not care about the course.

6. Tell your professor what you like about the course. Most students tell professors what is wrong with their course. Few let them know when they are doing things well. Set yourself apart from the rest. Highlighting the positive not only gives you an opening for conversation, but it also helps the professor prepare for future courses.

7. Telling the professor what you like about the course is particularly useful if you have a problem. Start with the positive comment and then request help for your problem. (I really enjoyed your discussion of .... but I could not read the red font on the slide.) Have a possible solution handy. (Could you use the black all the time?) Listen to what your professor is

saying. There may be a reason for what the professor is doing. (The red ink highlights key concepts) Be prepared to negotiate. (Perhaps you could write with a black pen and underline in red. Perhaps I could sit in the front row.) You will be more likely to solve the problem than if you simply complain.

8. If you make an appointment with a professor, keep it. If you absolutely cannot make it, cancel the appointment promptly. If you are caught in an emergency, call as soon as possible and explain the situation. Remember that your goal is to have your professor get a positive impression of you. If you do not show up you could be wasting a lot of time and appear not to take the course or the professor seriously.

9. Whenever you leave a message (voicemail, email, or written) be sure to include your name, course or section number and how and when your professor can contact you. Many students wonder why they never get a return call. A professor cannot call you if you do not leave a name, a working phone number or spoke to fast so that the phone number could not be understood. (You should also remember that everyone hears your answering machine message, not just your friends. You can seriously damage a positive impression if your professor has to listen to tasteless language, music, or jokes before the beep.)

10. Take advantage of email. You may get more information if you ask a question via email. Many professors find it easier to write a detailed response at a time when there are fewer interruptions. When you see the professor in person, remind him of your electronic communications. You can do this by directly referring to it or indirectly by thanking him for a rapid or helpful response.

11. In general, be friendly and courteous. Your professors are people too. They have good days and bad days. There are times when they are free to talk. There are other times when they have students waiting, a lecture to prepare, a meeting to attend, or a problem in the lab. Because of this, you might not always get the time and attention you feel you deserve. Make an appointment. Come back. Try again. Patience and persistence are usually rewarded.

III. Study Skills

A. Note taking

1. Add to diagrams in the Power Points and textbook.

2. Get the key facts down from the lecture.

3. When home, supplement your notes from the textbook readings.

4. Do not be afraid to rewrite your notes so that you will completely understand the material.

5. Make index cards for definitions and facts. Carry them with you and take them out when you are waiting (on lines, for classes to start, etc.)

6. Use of tape recorders: Check with the instructor to see if this is permitted. If you do tape the lecture, you may find it useful to limit your note taking to items that are written on the board or on overheads. Devote all your attention to the lecture. When you listen to the tape, that is the time to take notes—you have heard the material once, so you will know what where the lecture was going and what is important.

7. Make a concept map.

B. Time Management

1. Come to all classes—get there on time and stay until the end.

2. Note on a calendar when all assignments are due as soon as they are given.

3. Block out set times to study. These blocks should be at least an hour in length. There is no rule as to how long the sessions should be; but take a 5 to 10 minute break each hour. Do not overdo—when the material all becomes a blur it is time to stop for awhile.

4. Do not cram! Keep up with the material. Learning the material on a weekly basis, when it is test time all you will need to do is review. Learning as you go will also allow you to ask questions in lecture about material you do not understand.

C. *Where to Study*

1. Study in a regular area at a regular time. This will help you focus. This place should be quiet so all your attention can be devoted to your work.

2. Do not get too comfortable—remember you are here to study.

3. Do not to relax or fall asleep. It is best to work at a desk or table with a straight-backed chair.

4. Use the library.

5. Study Groups are useful if it is used to review and explain material and test the members. Do not fall into the trap of using it to socialize.

6. Take advantage of office hours to ask the instructor questions. Attend study sessions. Check with the Tutoring Center to see if BIOL 110 tutors are available.

IV. How to Study

1. **Survey** what you are to study. Read the following: objectives, titles, main headings, bold faced and italics words, diagrams and charts.

2. **Question** Turn the "Survey" into questions—ask yourself possible test questions.

3. **Read** the text carefully. Look for answers to your questions. After you read a section, highlight key words or phrases (no more than 10% of the material). Check your notes and update them where necessary.

4. **Recite** important points (silently, out loud or write). This forces your mind to remember.

5. **Review** the previous material before you start new material.

V. Memory Techniques

1. **Senses** Read the material, write, speak, visualize it

2. **Memorize key words** in order to remember complex concepts and processes.

3. **Use catchwords, acronyms and catch phrases** that represent the first letter of each keyword.

4. **Create rhymes** or song jingles

5. Create **visual** associations

6. **Break** large words into smaller ones

7. Study in small chunks instead of all at once.

8. **Reward** yourself

VI. Test Taking Skills

A. General

1. Review the night before.

2. Bring in all appropriate supplies to the test.

3. Arrive early to get organized—mentally and physically. Take a few "deep cleansing" breaths—this will help clear your mind and calm you down.

4. Directions: listen to verbal and carefully read written ones.

5. Watch the numbering. Make sure your answer to a question goes into the appropriate answer space.

B. Budget Your Time

1. Answer the easiest questions first. This builds confidence and will build associations for harder questions.

2. If you are stuck, go on. Later questions may trigger a response for the "stuck" one.

3. Do not look for a trick. Questions are usually straight forward. Do not ignore the obvious.

4. Reason out the tough questions:
   a. look for clues and "memory joggers" in the question or other questions
   b. look for clues in the answer choices
   c. eliminate obvious wrong choices

5. Usually your first instinct is right; but if you're are absolutely sure, change the answer

Time Line of Important Biological Discoveries

1660's Anton van Leeuwenhoek
Development of light microscopes

1665 Robert Hook
Coins the term cell

1838-39 Mathias Schleiden, Theodor Schwann
All organisms are composed of at least one cell

1859 Charles Darwin
Writes "On the Origin of the Species"

1864 Louis Pasteur
Biogenesis experiments with swan neck flasks

1866 Gregor Mendel
Publishes "Experiments with Plant Hybrids"

1868 George Huntington
Described the genetic disorder that is now named after him

Walther Flemming

1869 Friedrich Miescher
Discovers the molecule, DNA, in the nucleus and calls it nuclein

Discovery of chromosomes and mitosis

1880's Discovery mitosis and meiosis

1888 Term chromosome is first used

1898 Camillo Golgi
Describes the apparatus that bears his name

1900 Hugo deVries, Carl Correns, Erich deVries
Independently rediscover Gregor Mendel's work

1901 Karl Landsteiner

Human blood groups (ABO system)

Nobel Prize: 1930

Walter Sutton

Observes meiotic segregation of chromosomes and observes it matches Mendel's inheritance patterns

Archibald Garrod

Some diseases are inherited, thus Mendel's factors were controlling cellular chemistry

1904 William Bateson

Shows that some characteristics are not independently inherited, leading to the concept of "gene linkage"

Edmond Wilson and Nellie Stevens

Separate X and Y chromosomes determine sex

1907 Ernest Rutherford

Nuclear nature of atoms

Nobel Prize: 1908

1908 Archibald Garrod

Relationship between genes and enzyme production

1909 W. Johansen

Coins the word gene for Mendel's factors

1912 Thomas Hunt Morgan

Genes are on chromosomes and their role in hereditary

Nobel Prize: 1933

Theodore Richards

Accurately determined the atomic masses of over thirty important elements

Nobel Prize: 1914

Richard Willst•ter

Investigations into plant pigments, especially chlorophyll

Nobel Prize: 1915

1914 Arthur Harden, Hans von Euler-Chelpin

Alcoholic fermentation and fermentative enzymes

Nobel Prize: 1929

1915 Otto Meyerhof

Relationship between oxygen consumption and the metabolism of lactic acid in the muscle

Nobel Prize: 1922

Calvin B. Bridges

Discovery of non-disjunction

1919 Francis Aston

Discovery of isotopes of non-radioactive elements and the enunciation of the whole number rule for atomic mass

Nobel Prize: 1922

1927 Hermann Muller

Discovery of the production of mutations by means of X-ray irradiation

Nobel Prize: 1946

1928 Frederick Griffith

Bacterial transformation experiments

Emil Heitz

Described heterochromatin

1934 Axel Theorell

Discovery of the nature and mode of the action of oxidative enzymes (electron transport systems)

Nobel Prize: 1955

1937 Hans Krebs

Discovery of the metabolic (Citric Acid) cycle

Nobel Prize: 1953

Max Ferdinand Perutz, John Cowdery Kendrew

Structure of globular proteins, especially hemoglobin

Nobel Prize: 1962

1939 Linus Pauling

The nature of the chemical bond and the structure of molecules and crystals

Nobel Prize: 1954; only person to win the prize in two different disciplines—1963 Peace Prize

George Beadle and Edward Tatum

Develop the "one-gene one-enzyme' hypothesis

Nobel Prize: 1958

1943 Oswald Avery, Colin MacLeod, Maclyn McCarty
DNA is the transforming (genetic) material

Frederick Sanger
Structure of proteins, especially insulin
Nobel Prize: 1958

William Astbury
DNA has a regular, periodic pattern

1945 Melvin Calvin
Carbon dioxide assimilation in plants
Nobel Prize: 1961

Fritz Lipmann
Discovery of co-enzyme A and its importance for intermediary metabolism
Nobel Prize: 1953

1946 Charlotte Auerback, A. J. Clark and J. M. Robson
Chemicals can cause mutations

1947 Barbara McClintock
Transposons (jumping genes)
Nobel Prize: 1983

R. Briggs and T.J. King
Transfer of frog cell intestinal nuclei produced into denucleated eggs normal tadpoles

1952 Alfred Hershey and Margaret Chase
"Blender Experiments" show that DNA is the genetic material of phages

1953 Francis Crick and James Watson
Molecular structure of DNA
Nobel Prize: 1962

Jo Hin-Tijio
Humans have 46 chromosomes

1955 Christian deDuve
Discovered lysosomes
Nobel Prize: 1974

1956 Mohlon Hoagland and Paul Zamecnik
Discover tRNA

1957 Arthur Kornberg

Discovery of the first DNA polymerase enzyme

Nobel Prize: 1959

Francis Crick and George Gamov

Work out the "Central Dogma" of Biology

1958 Matthew Meselson, Franklin Stahl

DNA replication is semi-conservative

Jerome Lejeune

Downs Syndrome is caused by a trisomy of chromosome 21

John Gurdon

Cloned a frog using intact nuclei from somatic cells of a tadpole

Nobel Prize: 2012

François Jacob, Jacques Monad and André Lwoff

Cells can turn on and off certain genes

Nobel Prize: 1965

1960 Sydney Brenner, Francis Crick, François Jacob and Jacques Monod

Discovery of mRNA

1961 Peter Mitchell

Chemiosmotic phosphorylation of ATP

Nobel Prize: 1978

Daniel Nathans, Hamilton Smith, Werner Arbel

Restriction enzymes and their application to problems of molecular genetics

Nobel Prize: 1978

Mary Lyon

Mechanism for inactivation of one of the X chromosomes in mammalian females

1962 Chamberlin Berg

Discovery of tRNA

late 1960's Reiji and Tuneko Okazaki

Discontinuous mode of DNA replication

1965 Christian deDuve

Discovered peroxisomes

1966 **Robert Holley, Har Khorana, Marshall Nirenberg**

Interpretation of the genetic code and its function in protein synthesis

Nobel Prize: 1968

1967 **Lynn Margulis**

First article on the Endosymbiotic Theory

Francis Crick

Advances the Central Dogma of Biology

Howard Temin, David Baltimore

Reverse transcriptase

Nobel Prize: 1975

1970 **Howard Tomin and David Baltimore**

Independently discover reverse transcriptase

Hamilton O. Smith

Site specific restriction enzyme

1972 **S.J. Singer, G.L. Nicolson**

Fluid Mosaic Model of membrane structure

Harry Noller

Role for rRNA in the translation of mRNA into protein

Paul Berg

Creates first recombinant DNA molecule

Michael S. Brown, Joseph L. Goldstein

Regulation of cholesterol metabolism (receptor mediated endocytosis)

Nobel Prize: 1985

Janet D. Rowleg

Discovered the process of translocation

First recombinant DNA organism

1975 **E.M. Southern**

Procedure to identify the locations of genes and other DNA sequences on restriction fragments by gel electrophoresis (Southern Blot)

Richard Roberts and Phil Sharp

Discovery of exon and intron regions in DNA

Dawn of Biotechnology: human protein is made in a bacteria

1977 Carl R. Woese

Used molecular structures, especially rRNA, to develop a classification system for living organisms that is based on three domains

Walter Gilbert and Frederick Sanger

Techniques to sequence DNA

1978 Paul Boyer, John Walker, Jens Skore

Enzymatic mechanisms underlying the synthesis of ATP

Nobel Prize: 1997

Diamond vs. Charkabarty

Supreme Court allows patent for an oil dissolving microbe

David Botstein

Use of restriction fragment length polymorphism (RFLP) in gene mapping to indicate difference between individuals

1980s Sidney Altman and Thomas Cech

RNA (ribozymes) may function as a catalyst

Nobel Prize: 1989

1981 1st transgenic animal

Human insulin is the first successful pharmaceutical product of recombinant DNA technology

Genetic marker for Huntington's is found

1983 Johann Deisenhofer, Robert Huber, Hartmut Michel

Determination of the three dimensional structure of a photosynthetic reaction center

Nobel Prize: 1988

1984 Kary Mullis

Polymerase Chain Reaction (PCR)

Nobel Prize: 1993

Elizabeth Blackburn, Carol Greider, Jack Szostak

Discovered telomeres and telomerase

Nobel Prize: 2009

DNA "fingerprints" used in a forensic case

1986 Leroy Hood

Automated sequencer

1989 James Watson

Creates the National Center for Human Genome Research to decode the human genome

First attempt at "curing" a human, genetic disorder

1991 J. Craig Venter

Fast approach to gene mapping using Expressed Sequence Tags (ESTs)

1992 Peter Agre

Discovered aquaporins

Nobel Prize: 2003

1990's FlavrSavr™: 1st genetically modified food approved by FDA for sale

Ban on genetic discrimination in the workplace

1996 Ian Wilmut

Dolly, cloned sheep, using early sheep embryo nuclei

2003 Human Genome Project completed

2007 Shinya Yamanaka

IPS cells: mature cells can be reprogrammed to become pleuripotent stem cells

Nobel Prize: 2012

UNIT 1: CHEMISTRY

Outline

TOPIC	SOLOMON	SACK.
I. Inorganic Chemistry	26-27	
A. Atomic Structure		3-7
1. The Atom		
2. Atoms vs. Elements		
3. Molecules vs. Compounds		
4. Compounds vs. Mixtures		
B. Using the Periodic Table	28-29	
1. "Normal" Atoms		19-23
a. Atomic Number		7
b. Atomic Mass		8-10
2. Isotopes		10-12
C. Electrons and Energy Levels	28-29	13-18
D. Ions		29-41
1. Electron Movement		
2. Valence		
3. Bonds		
a. Ionic	33-34	42
b. Covalent	31-33	101-114
(i). Polar vs. Non-Polar		121-125
c. Electrolytes vs. Non-electrolytes		46-47
d. Hydrogen Bonds	35	151-126
E. Properties of Water	36-39	
F. Specific Molecules	39-41	
1. Acids, bases and salts		43-46
2. pH		49-54
3. Buffers		55-58
G. Equations		
1. Molecules	30	25-27
2. Balanced Equations	31	
a. Oxidation vs. Reduction	35-36	241-244
II. Organic Chemistry		
A. Definition	45	
B. Functional Groups	47-48	
1. Definition		
2. Hydrocarbons		
3. Types		
a. Alcohol (hydroxyl)		127-129
b. Carbonyls		

TOPIC	SOLOMON	SACK.
(1) Aldehyde		130-135
(2) Ketone		130-135
(3) Esters		
c. Acid (carboxyl)		136-137
d. Amine		138-141
III. Biochemistry		
A. Monomeres and Polymers	47	
B. Formulas	30	
1. Structural Isomers	46	155-160
C. Reactions	48	
1. Dehydration Synthesis/Condensation		
2. Hydrolysis		
D. Carbohydrates	49-54	171-180
1. Monosaccharides		
2. Disaccharides		
3. Polysaccharides		
4. Nutrition		
E. Lipids	54-58	
1. Triglycerides (neutral fats)		181-184
2. Phospholipids		185
3. Carotenoids		
4. Waxes		
5. Steroids		186-187
6. Nutrition		
F. Proteins	58-66	
1. Structure		193-197
2. Formation		189-192
a. Levels of Organization		
3. Denature		198-199
4. Nutrition		
IV. Sackheim Review		146-150, 289-309
V. Supplements		
A. Thinkwell CD-ROM		
1. Inorganic and Organic Chemistry		
a. An Introduction to Atoms		
b. Atoms and Bonding		
c. Properties of Water		
d. Water Quiz and Review		
e. Carbon Chemistry		
f. Carbohydrates		
g. Lipids (and Nucleic Acids)		
h. Proteins		
VI. Exam		

Objectives

1. Define the terms: matter, atom, element, molecule, compound, mixture, chemical and physical change.

2. Identify the subatomic particles, their location, charges and masses.

3. Given the Periodic Table or the element written in proper form, for the first 20 elements:
   a. identify their symbols.
   b. identify their atomic numbers.
   c. identify their atomic masses.
   d. determine their number of protons, neutrons and electrons for any isotope.
   e. determine their number of energy shells and number of electrons in each shell.
   f. determine their valence number.

4. Define the term ion and discuss how it is formed.

5. Differentiate between:
   a. oxidation and reduction. Recognize them in equation form.
   b. electrolyte and non-electrolyte.

6. Be able to determine:
   a. the number of atoms in a molecule.
   b. the type of bonding that is present.

7. For balanced equations, determine the reactants and products.

8. Discuss the differences between acids, bases and salts and their relationship to pH. Discuss the role of buffers.

9. Explain the formation of covalent bonds and relate to organic molecules.

10. Explain how the electro-negativity of atoms within a molecules influences the polarity of the molecule. Show how the molecule is hydrophobic or hydrophilic.

11. Describe hydrogen bonds and explain how they are formed.

12. Contrast the relative strengths of ionic, covalent and hydrogen bonds.

13. Describe the properties of water and explain these properties in terms of the chemistry of the water molecule.

14. Examine structural and/or molecular formulas and identify the various organic functional groups that are listed in the outline.

15. Define the term isomer. Recognize structural isomers and state the isomers of glucose.

16. Differentiate between dehydration synthesis and hydrolysis.

17. For carbohydrates, lipids and proteins, discuss the following:
    a. chemical and structural characteristics.
    b. types of subunits, bonding and classification scheme.
    c. examples of these macromolecules.
    d. food sources.
    e. importance to cells and organisms.

Worksheet 1.1: The Atom

Element	Symbol	Atomic #	# Protons	# Electrons	Mass #	# Neutrons
Carbon						
Carbon-14						
Chlorine						
Chlorine-37						
Iron						

Worksheet 1.2: Ions

Atom	# electrons	1st Orbital	2nd Orbital	3rd Orbital	4th Orbital	Gain/Lose	Valence	Ion
Li								
Al								
O								
Ne								
C								
Cl								
N								
Ca								

Figure 1.1: Oxidation

The gain of oxygen

```
       H  H                  H  OH
       |  |                  |  |
    H—C—C=O →             H—C—C=O
       |                     |
       H                     H
```

The loss of hydrogen

```
  COOH          COOH
  |             |
 H-C-OH    →    C=O
  |             |
  COOH          COOH
```

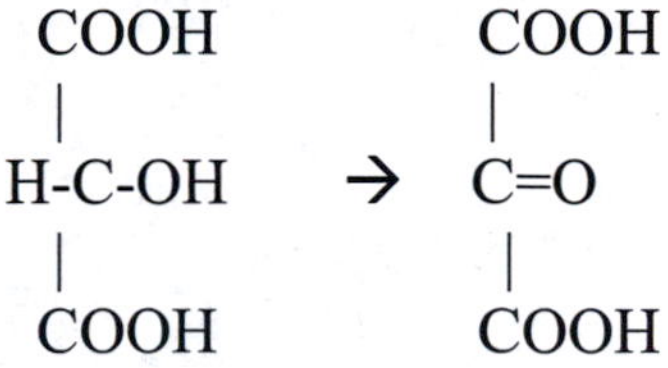

Figure 1.2: Reduction

The loss of oxygen

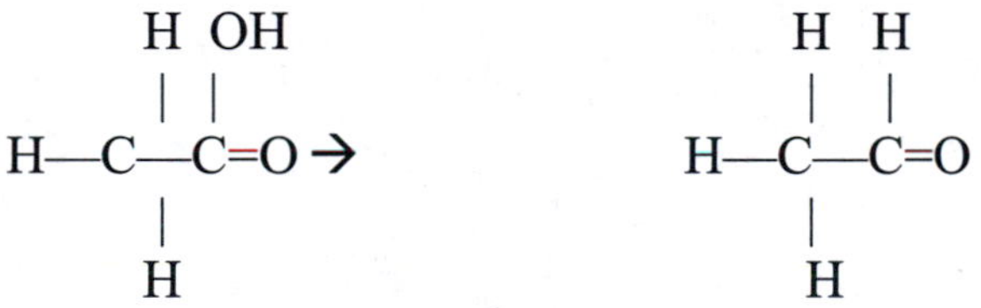

```
       H  OH                 H  H
       |  |                  |  |
    H—C—C=O →             H—C—C=O
       |                     |
       H                     H
```

The gain of hydrogen

```
  COOH          COOH
  |             |
  C=O     →   H-C-OH
  |             |
  COOH          COOH
```

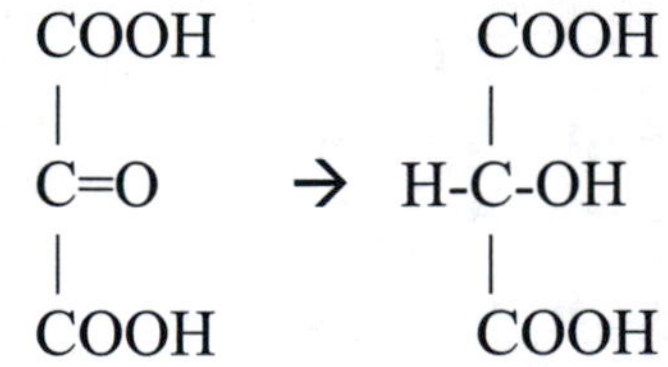

Figure 1.3 Organic Functional Groups

1. Hydrocarbons: carbon and hydrogen molecules

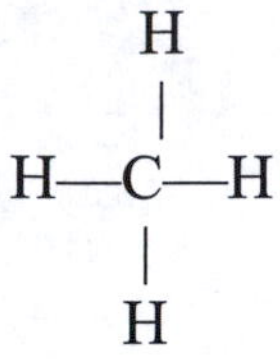

H
|
H—C—H = CH_4
|
H

Methane

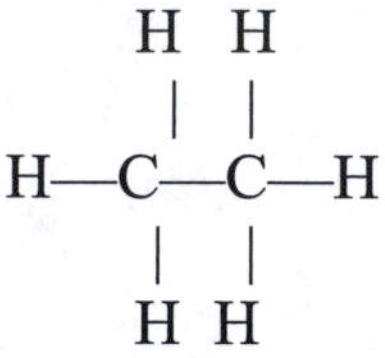

H H
| |
H—C—C—H = C_2H_6
| |
H H

Ethane

2. Alcohols: R—OH; words end in -ol or alcohol

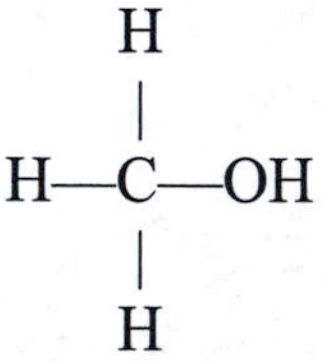

H
|
H—C—OH = CH_3OH
|
H

Methanol (Methyalcohol)

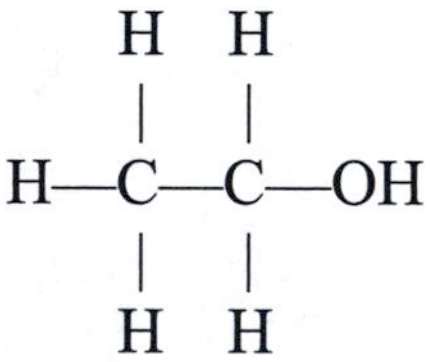

H H
| |
H—C—C—OH = C_2H_5OH
| |
H H

Ethanol (Ethalalcohol)

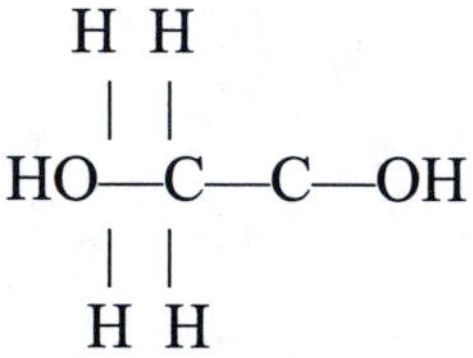

H H
| |
HO—C—C—OH = OHC_2H_4OH
| |
H H

Ethylene Glycol

$C_3H_5(OH)_3$ = *Glycerol*

3. Aldehydes: R—CH or CHO; words end in -al or aldehyde
‖
O

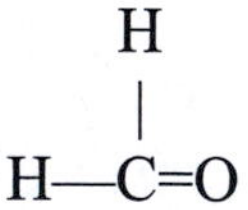

H
|
H—C=O = HCHO

Formaldehyde

4. Ketones: R—C—C

```
4. Ketones: R—C—C
              ||
              O
```

```
      H   O   H
      |   ||  |
  H—C—C—C—H
      |   |
      H   H
```

$= CH_3COCH_3$

Acetone (dimethyl ketone)

5. Acids: R—C==O or COOH or carboxyl group

```
5. Acids: R—C==O
            |
            OH
```

```
      H   O
      |   ||
  H—C—C—OH
      |
      H
```

$= CH_3COOH$

Acetic Acid

```
      H  H   O
      |  |   ||
  H—C—C—C—OH
      |  |
      H  OH
```

$= CH_3CHOHCOOH$

Lactic Acid

6. Esters: R—C—O—R or COOC

```
6. Esters: R—C—O—R
             ||
             O
```

```
      H  O        H  H
      |  ||       |  |
  H—C—C—O—C—C—H
      |           |  |
      H           H  H
```

$= CH_3COOC_2H_5$

Ethyl Acetate

7. Amines: R—N—H or NH_2

```
7. Amines: R—N—H
             |
             H
```

Figure 1.4: Isomers

Same molecular formula: C_5H_{12}

Different structural formulas:

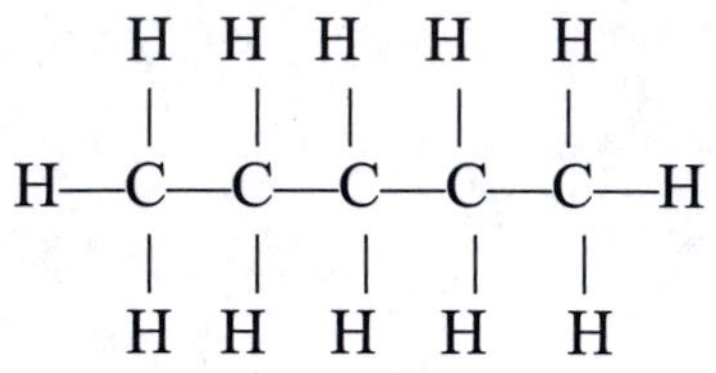

```
      H  H  H  H  H                  H  H  H  H
      |  |  |  |  |                  |  |  |  |
   H—C—C—C—C—C—H        OR       H—C—C—C—C—H
      |  |  |  |  |                  |  |  |  |
      H  H  H  H  H                  H  |  H  H
                                      H—C—H
                                         |
                                         H
```

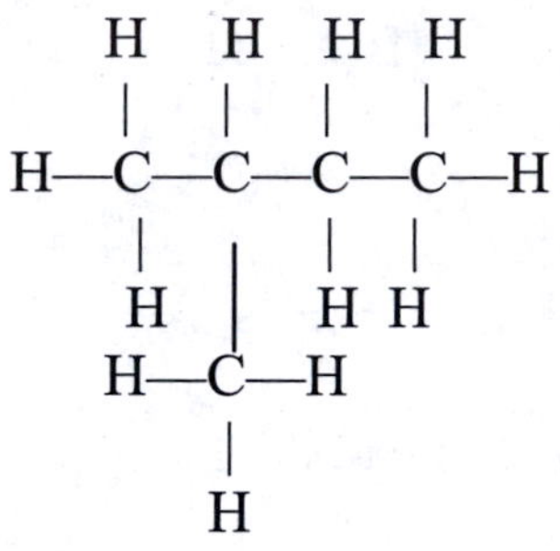

Figure 1.5: Biochemistry

MOLECULE	FOOD SOURCES	CHEMICAL CHARACTERISTICS	FUNCTION IN BODY
CARBOHYDRATES		Hydrophilic	
Monosaccharides	Sweet fruits, vegetables	C, H, O ratio is 1:2:1 , minimum 4 carbons	Principle energy source
Disaccharides	Honey, sugar sap, milk	C, H, O ratio is 12:22:11	Transported in plants and animals to be stored as polysaccharides or broken down to monosaccharides
Polysaccharides	Seeds of wheat, corn & rice, bread, potatoes	$(C_6H_{10}O_5)_n$	Storage of carbohydrates
LIPIDS		Hydrophobic	
Triglycerides/Neutral Fats		Glycerol & 3 Fatty Acids	Energy reserves. Insulates against heat loss. Protect vital organs. Absorption of fat soluble vitamins
Saturated Fatty Acids	Meats	No carbon to carbon double bonds; solid at room temperature	
Unsaturated Fatty Acids	Vegetables	At least one carbon to carbon double bond; liquid at room temperature	
Phospholipids		Glycerol, 1 saturated fatty acid, 1 unsaturated fatty acid, phosphate group	Membranes
Waxes		Long carbon fatty acid and long carbon alcohol	Protection: animal fur, outer ear. Plants: coats leaves for water retention
Carotenoids	Yellow & orange plant pigments	5 carbon isoprene monomers	Converted to Vitamin A
Steroids	Eggs, milk, cheese, shellfish	4 interlocking carbon rings	Vitamins A, D, E, K; sex hormones; bile salts; cholesterol

MOLECULE	FOOD SOURCES	CHEMICAL CHARACTERISTICS	FUNCTION IN BODY
PROTEINS		20 amino acids	Structures (hair, collagen, etc.), hormones, antibodies, enzymes, transport molecules across membranes, recognition
		Primary Structure: sequence of amino acids to form polypeptide chain Secondary Structure: alpha helix/coil and beta sheet Tertiary Structure: folding to form globular structure; di-sulfide bridges Quaternary: joining of 2 or more tertiary structures	
Essential Amino Acids (cannot be manufactured by the body)	Meat, poultry, fish, eggs	Complete proteins: contains all 8 essential amino acids	
	Nuts, grains, vegetables, legumes	Incomplete proteins: missing at least 1 essential amino acid	

After Class Work

1. From the given information, fill in the rest of the chart

NAME	SYMBOL	ATOMIC NUMBER	ATOMIC MASS	NUMBER PROTONS	NUMBER NEUTRONS	NUMBER ELECTRONS
		20				
	Be					
		8				
				15		
				7		
	C-14					
	Ar					
				2		
Lithium						
				10		
Aluminum						
	Si					
	$_5B^{15}$					

2. From the given information fill in the rest of the chart

NAME	SYMBOL	ATOMIC NUMBER	ATOMIC MASS	NUMBER PROTONS	NUMBER NEUTRONS	NUMBER ELECTRONS
Chlorine Ion						
	Mg^{+2}					
Fluorine Ion						
	S^{-2}					
Sodium Ion						

3. Identify the number of electrons in the outer shell for each of the following elements:

a. O____________________ f. C____________________

b. H____________________ g. F____________________

c. Cl____________________ h. S____________________

d. N____________________ i. Al____________________

e. Ne____________________ j. P____________________

4. Identify the valence for each of the following elements:

a. Cl________________________ f. S________________________

b. Na________________________ g. F________________________

c. Mg________________________ h. Al________________________

d. O________________________ i. N________________________

e. P________________________ j. Ca________________________

5. $Na + Cl \rightarrow Na^+ + Cl^-$

a. Is Na being reduced or oxidized?

b. Is Cl being reduced or oxidized?

$FAD + H_2 \rightarrow FADH_2$

c. Is FAD being reduced or oxidized?

6. What type of reaction is shown? Which molecule is oxidized? Which molecule is reduced?

a.

```
   COOH                      COOH
    |                         |
 H—C—H                       C—H
    |            →           ||
 H—C—H                       C—H
    |                         |
   COOH                      COOH
```

Succinic Acid   Fumaric Acid

b.

```
  H                H
  |                |
 C=O    →      H—C—OH
  |                |
 CH3              CH3
```

Acetaldehyde   Ethanol

7. Write the name & numbers of all the elements found in the molecules listed.

a. NH_3

b. HCl

c. H_2

d. $Ca(OH)_2$

e. Na_2CO_3

f. N_2O_5

g. $C_6H_{12}O_6$

h. H_2SO_4

i. NH_4Cl

j. $CH_3(CH_2)_3COOH$

8. For each reaction, name the products and reactants.

a. $Mg + O_2 \rightarrow MgO$

b. $Zn + 2HCl \rightarrow ZnCl_2 + H_2$

c. $C + O_2 \rightarrow CO_2$

d. $3Fe + 2O_2 \rightarrow Fe_3O_4$

e. $2NaCl + AgNO_3 \rightarrow AgCl + 2NaNO_3$

f. $Mg + 2AgNO_3 \rightarrow Mg(NO_3)_2 + 2Ag$

g. $ZnSO_4 + 2NaOH \rightarrow Zn(OH)_2 + Na_2SO_4$

9. Tell if the answer is an acid, base, salt or buffer.

a. ________, ________, ________ are substances that ionize in water and are good electrolytes

b. ________ are proton (H^+) acceptors

c. ________ ionize in water to release hydrogen ions and a negative ion other than hydroxide (OH^-)

d. ________ ionize in water to release ions other than H^+ and OH^-

e. ________ are formed when an acid and a base combine

f. ________ include substances such as lemon juice and vinegar

g. ________ prevent rapid/large sings in pH

h. ________ pH <7

i. ________ pH >7

j. ________ when added to a solution will not change the pH

10. Tell if the statement applies to an acid, base or neutral solution.

a. ________ more H^+ ions than OH^- ions

b. ________ more OH^- ions than H^+ ions

c. ________ equivalent number of H^+ and OH^- ions

d. ________ pH = 3

e. ________ pH = 7

f. ________ pH = 11

11. Classify each of the following molecules as either polar or non-polar covalent bonding and explain whether they are hydrophobic or hydrophilic.

a.

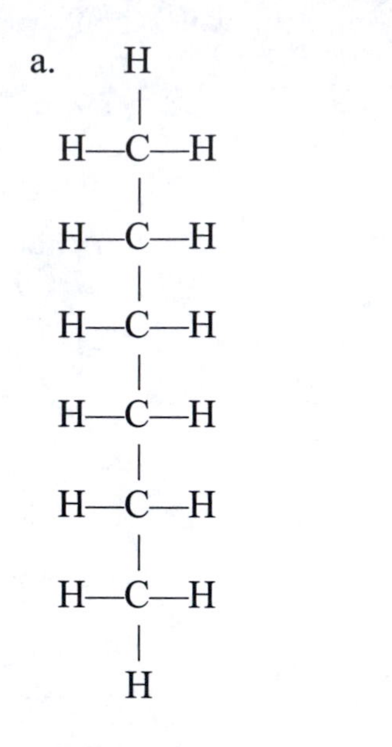

b.

```
   H
   |
   C=O
   |
H—C—O—H
   |
H—C—O—H
   |
H—C—O—H
   |
H—C—O—H
   |
H—C—O—H
   |
   H
```

12. Compare the following 2 molecules and determine in what ways they are similar and how they are different.

a.

```
      H
      |
      C=O
      |
   H—C—O—H
      |
H—O—C—H
      |
   H—C—O—H
      |
   H—C—O—H
      |
   H—C—O—H
      |
      H
```

b.

```
         H
         |
      H—C—O—H
         |
         C=O
         |
   H—O—C—H
         |
   H—O—C—H
         |
   H—O—C—H
         |
   H—O—C—H
         |
         H
```

13. Which of the following is saturated and which is unsaturated?

a.
```
    H  H  H
    |  |  |
H—C=C—C—C=O
          |  |
          H  OH
```

b.
```
    H  H  H
    |  |  |
H—C—C—C—C=O
    |  |  |  |
    H  H  H  OH
```

15. a. Complete the following equation.

```
H    H O                  H   H  O
|    |  ||                |   |  ||
H—N—C—C        +        H—N—C—C
     |  |                     |  |
     R1 OH                    R2 OH
```

b. What is the name for what has taken place?

c. What is the name for the type of bond that has formed?

d. What is the name for the reverse of this reaction?

Answers

1.

NAME	SYMBOL	ATOMIC NUMBER	ATOMIC MASS	NUMBER PROTONS	NUMBER NEUTRONS	NUMBER ELECTRONS
Calcium	Ca	20	40	20	20	20
Beryllium	Be	4	9	4	5	4
Oxygen	O	8	16	8	8	8
Phosphorous	P	15	31	15	16	15
Nitrogen	N	7	14	7	7	7
Carbon 14	C-14	6	14	6	8	6
Argon	Ar	18	40	18	22	18
Helium	He	2	4	2	2	2
Lithium	Li	3	7	3	4	3
Neon	Ne	10	20	10	10	10
Aluminum	Al	13	27	13	14	13
Silicon	Si	14	28	14	14	14
Boron-15	$_5B^{15}$	5	15	5	10	5

2.

NAME	SYMBOL	ATOMIC NUMBER	ATOMIC MASS	NUMBER PROTONS	NUMBER NEUTRONS	NUMBER ELECTRONS
Chlorine Ion	Cl^{-1}	17	35	17	18	18
Magnesium Ion	Mg^{+2}	12	24	12	12	10
Fluorine Ion	F^{-1}	9	19	9	10	10
Sulfur Ion	S^{-2}	16	32	16	16	18
Sodium Ion	Na^{+1}	11	23	11	12	10

3. a. 6
   b. 1
   c. 7
   d. 5
   e. 8
   f. 4
   g. 7
   h. 6
   i. 3
   j. 5

4. a. -1
   b. +1
   c. +2
   d. -2
   e. -3
   f. -2
   g. -1
   h. +3
   i. -3
   j. +2

5. Oxidized or reduced is the state the molecule is in (with or without e's of H's). Oxidation or reduction is the reaction that removes or adds e's of H's.
   a. being oxidized because it is losing an electron
   b. being reduced because it is gaining an electron
   c. FAD is being reduced because it is gaining hydrogen

6. a. This is a oxidation reaction because hydrogens are lost. The succinic acid is reduced because it has more hydrogens than fumaric acid which is oxidized.
b. The reaction is reduction because hydrogens are gained. Acetaldehyde is oxidized because it has less hydrogens than ethanol which is reduced.

7. a. nitrogen-1, hydrogen-3
   b. hydrogen-1, chlorine-1
   c. hydrogen-2
   d. calcium-1, oxygen-2, hydrogen-2
   e. sodium-2, carbon-1, oxygen-3
   f. nitrogen-2, oxygen-5
   g. carbon-6, hydrogen-12, oxygen-6
   h. hydrogen-2, sulfur-1, oxygen-4
   i. nitrogen-1, hydrogen-4, chlorine-1
   j. carbon-5, hydrogen-10, oxygen-2

8. If every reaction, the reactants are to the left of the arrow and products are to the right.

Reactants → Products

9. a. acid, base, salt
   b. base
   c. acid
   d. salt
   e. salt
   f. acid
   g. buffers
   h. acid
   i. base
   j. salt

10. a. acid
    b. base
    c. neutral
    d. (strong) acid
    e. neutral
    f. (strong) base

11. a. Non-polar because the electrons are equally shared between the carbons and hydrogens. Because it is non-polar, it has no charges distributed along the molecule; therefore it is hydrophobic—will not dissolve in water. There are six carbons and no oxygens and no nitrogens thus fitting the general formula for non-polar.

b. Polar because the electrons are not equally shared between the carbons and oxygens. Oxygen has 2 unpaired electrons, therefore electrons are spending more time with oxygen than with carbon; thus making oxygen partially electro-negative. The partial charges allow it to be hydrophilic--will be pulled between the charges on the water molecule. There are six carbons and six oxygens and no nitrogens thus fitting the general formula for polar.

12. These molecules have the same number and type of atoms (same formula: $C_6H_{12}O_6$) but have a different arrangement of the atoms (structure). They are called isomers of each other.

13. Molecule (a) is unsaturated because it contains at least 1 double bond between the carbons (more hydrogens can enter the molecule). Molecule (b) is saturated because it contains all single bonds and no more hydrogens may enter the molecule.

14. a.

```
    H   H  O   H    H  O
    |   |  ||  |    |  ||
H—N—C—C—N—C—C
        |           |   |
        R1          R2  OH
```

b. dehydration synthesis
c. peptide bond: a specific type of covalent bond
d. hydrolysis

Sample Multiple Choice Questions

1. An atom has six protons and eight neutrons. Its atomic mass is _____ atomic mass units.
    a. two
    b. four
    c. six
    d. eight
    e. fourteen

2. Isotopes differ from each other with respect to the number of
    a. protons.
    b. electrons.
    c. neutrons.
    d. protons and electrons.
    e. neutrons and protons.

3. In the formation of common table salt, sodium and chlorine interact because
    a. sodium and chlorine share a pair of electrons.
    b. sodium and chlorine share two pairs of electrons.
    c. chlorine donates seven electrons to sodium.
    d. sodium donates one electron to chlorine.

4. The cohesiveness between water molecules is due largely to
    a. hydrogen bonds.
    b. polar bonds.
    c. non-polar bonds.
    d. ionic bonds.
    e. hydrophobic interactions.

5. A pH of 4 is _____ times more _____ than a pH of 7.
    a. a; basic
    b. 3; acidic
    c. 1000; neutral
    d. 1000; basic
    e. 1000; acidic

6. Which of the following is *not* a property of carbon?
    a. Carbon to carbon bonds are limited to single bonds.
    b. Carbon has 4 valence electrons.
    c. Carbon can form bonds to various other atoms.
    d. Two carbon atoms can share three electron pairs with each other.
    e. Carbon to carbon bonds are very strong.

7. Glucose dissolves in water because it
    a. ionizes.
    b. is a polysaccharide
    c. has polar hydroxyl groups that interact with polar molecules.
    d. has a very reactive primary structure.
    e. is hydrophobic.

8. A chemical reaction in which organic molecules are made from smaller subunits is called
    a. hydrolysis
    b. condensation.
    c. oxidation.
    d. reduction.
    e. neutralization.

9. Which of the following molecules is *not* grouped with the lipids?
    a. Waxes
    b. Steroids
    c. Cholesterol
    d. Carotenoids
    e. None of the above, all are lipids.

10. The tertiary structure of proteins is typified by the
    a. association of several polypeptide chains by weak links.
    b. order in which amino acids are joined in a peptide chain.
    c. bonding of two amino acids to form a dipeptide.
    d. folding of a peptide chain to form an alpha helix.
    e. three-dimensional shape of an individual polypeptide chain.

Answers to Multiple Choice Questions

1. E. The mass of an atom is determined by the only two sub-atomic particles that contain an appreciable mass—the protons and neutrons. Each of these sub-atomic particles has a mass of 1 amu, so you just add the number of protons to the number of neutrons. For this problem, 6 (protons) + 8 (neutrons) will give you a mass of 14. (The number of protons tells you the atomic number. For this problem the atomic number is 6. Since it is an atom, the number of protons must equal the number of electrons.)

2. C. An isotope is a variation of an atom. Since it is the same atom, the number of protons will remain the same. What accounts for the variation is the difference is mass and since you may never change the number of protons in an atom the number of neutrons will vary.

3. D. In class, table salt was used as an example of an ionic bond. Sodium has 1 electron in its valence (outer) shell and chlorine has 7 valence electrons. Therefore sodium will donate 1 electron and chlorine will accept one and an ionic bond will form.

4. A. Cohesiveness means a water molecule sticks to another water molecule. Since you are "joining" two molecules together, it must be a hydrogen bond that forms the attraction. Polar covalent bonds cause hydrogen bonds, but the polar bonds are not what is holding the water molecules together.

5. E. The pH scale is logarithmic, therefore the difference between each number is ten fold. To go from 4 to 7 you "travel" from 4 to 5 (10X), 5 to 6 (10X) and 6 to 7 (10X). To calculate the difference between 4 and 7 you multiply 10 times 10 times 10 and get 1,000. pH 7 is neutral, <7 is acidic and >7 is basic/alkaline.

6. A. Carbon can form single, double and triple covalent bonds. Choice D is a description of how a triple bond is formed.

7. C. Glucose is a monosaccharide belonging to the class of macromolecules called carbohydrates. In carbohydrates, for every carbon there is generally an oxygen. Thus the molecule is polar (unequal sharing of electrons between these atoms). Water is also a polar molecule so other polar molecules can be dissolved in water (hydrophilic).
Why A is wrong: Because glucose is composed of covalently bonded atoms, when it enters the water it remains in its molecular formation; unlike ionically bonded compounds which break down into ions.
Why D is wrong: Primary structure refers to proteins.
Why E is wrong: Hydrophobic (created by non-polar covalent bonds) means will not dissolve in water.

8. B. Is another name for dehydration synthesis. Choice A is the opposite—large molecules are broken down into smaller ones with the addition of water. Oxidation is the loss of hydrogens and reduction is the gain of hydrogens. A neutralization reaction is what occurs when an acid and base are mixed together to form a salt and water.

9. E. All of them are lipids because they have a high carbon content and a low number of oxygens. For completeness, waxes are made of a long, hydrocarbon alcohol and a fatty acid. Cholesterol is a steroid (4 inter-locking ringed structure) and carotenoids are lipids because they do not dissolve in water.

10. E. The primary structure of a protein is the sequence of the amino acids in the polypeptide chain. (Choices B and C) The secondary structure (Choice D) creates alpha coil/helixes and beta sheets. The tertiary structure creates the 3-dimensional structure called a globular protein. Some proteins, such as hemoglobin, have a quaternary structure where two or more polypeptide chains that are at the tertiary level join together.

UNIT 2: CELLS & MEMBRANES

Outline

TOPIC	SOLOMON
From Previous Units	
A. Macromolecules	
1. Cellulose	
2. Phospholipids	
3. Protein Structure	
B. Bonding Patterns	
1. Polar and Non-Polar	
2. Ionic	
II. What is Life	
A. Requirements for Life	1-10, 14-15
1. Characteristics	
2. Control Mechanisms	
a. Homeostasis	
b. Negative Feedback	
c. Positive Feedback	
B. Taxonomy	10-14
C. Domains and Kingdoms	
1. Food Requirements	
III. Cell Structure	
A. Cell Theory	72-74
1. What is a cell?	
B. Prokaryotic Cells	80
1. Cell Wall	
2. Flagella	
3. Cell Membrane	
4. Protoplasm	
5. Chromatin Material/Nucleoid	
6. Ribosome	
C. Eukaryotic Cells	81-100
1. Cell Size	
2. Cell Wall	
a. Extra-cellular Matrix	
3. Cell Membrane	
4. Flagella and Cilia	
5. Protoplasm vs. Cytoplasm	
6. Nucleus	
7. Nucleolus	

TOPIC	SOLOMON
8. Centriole	
9. Ribosome	
10. Endoplasmic Reticulum	
11. Vacuole and Vesicle	
a. Lysosome	
b. Perixosomes	
12. Golgi Complex	
13. Chloroplast	
14. Mitochondria	
16. Cytoskeletom	
a. Microfilaments/Actin	
b. Intermediate Filaments	
c. Microtubules	
D. Comparisons of Cell Types	
1. Prokaryotic vs. Eukaryotic	
2. Animals vs. Plant	
E. Endosymbiotic Theory	
IV. Cell Membrane	
A. Membrane Components	105-111
1. Fluid Mosaic Model	
B. Roles of the Membrane	
1. Cell Junctions: Anchoring & Communication	123-126
a. Spot Desmosome/Anchoring Junction	
b. Tight/Occulding Junction	
c. Gap Junction	
d. Plasmodesmata	
2. Recognition	
3. Transport Systems	112-113
a. Passive Transport	113-118
(i). Diffusion	
(ii). Osmosis	
(iii). Facilitated Diffusion	
(a). Channel Proteins	
(1). Aquaporins	
(2). Gated Channels	
b. Transporters	
(i). Uniporters	
(ii). Symporters	
(iii). Anti-porters	
c. Active Transport	118-122
(i). Carrier Mediated	
(a). Na^+/K^+ Pump/antiporter	
(b). H+ Pump	

TOPIC **SOLOMON**

(ii). Bulk Transport
(a). Endocytosis
(1). Phagocytosis
(2). Pinocytosis
(3). Receptor Mediated
(b). Exocytosis
C. Endomembrane System

V. Supplements
A. Tapes in Library
1. National Geographic: Discovering the Cell
2. Chemistry: The Cell and Energetics
a. Part 2: Journey into the Cell
b. Part 3: Endocytosis
c. Part 4: Cellular Secretion
3. Physiology
a. Part 2: Osmosis
b. Part 3: Active Transport Across a Cell Membrane
c. Part 4: Cellular Secretion
B. Thinkwell CD-ROM
1. Evolution
a. Classifying Life
2. Cell Biology
a. An Introduction to Cell Biology
b. Membrane Bound Organelles
c. Cytoskeleton
d. Plasma Membrane
e. Cell Transport

VI. Exam

Objectives

1. List, describe and give examples for the characteristics of living things. Define homeostasis and how an organism uses feedback mechanisms to maintain homeostasis.

2. Define evolution and explain the process of evolution by Natural Selection as proposed by Charles Darwin.

3. Explain how antibiotic resistance in bacteria demonstrates microevolution.

4. State the hierarchy of organization and relate to increasing complexity and energy requirements.

5. List, in proper sequence, the levels of taxonomic classification.

6. Name the current Domains and the imbedded Biological Kingdoms. Be able to give the characteristics and examples of each. Define autotroph, heterotroph and decomposer.

7. Explain what is meant by the unity and the diversity of life. Give examples.

8. Explain The Cell Theory and its importance.

9. State the structure and functions of the major components of prokaryotic cells. (See outline for listing.)

10. Using the terms total surface area and volume, explain why cells remain small.

11. State the structure and functions of the major components of eukaryotic cells. (See outline for listing.)

12. Identify the major similarities and differences between:
    a. prokaryotic and eukaryotic cells.
    b. eukaryotic animal and eukaryotic plant cells.
    c. similarities between all cells.

13. Explain the Endosymbiotic Theory and the available evidence regarding the origin of chloroplasts and mitochondria.

14. Describe the structure of the plasma membrane according to the Fluid-Mosaic Model and explain why the membrane is semi-permeable.

15. State three ways that animal cells and one way that plant cells are joined together. Explain how each mechanism functions.

16. Describe and explain the general phenomenon of diffusion. Indicate what molecules are able to diffuse across the plasma membrane.

17. Define osmosis and be able to apply to problem solving situations utilizing the terms isotonic, hypertonic and hypotonic.

18. Define facilitated transport/diffusion. Explain the roles of various transporter proteins in moving molecules across the membrane.

19. Define active transport.

20. Describe how pump systems operate. Explain why the Na^{+}/K^{+} Pump is an antiporter system.

21. Compare and contrast facilitated transport (diffusion) and the pump systems of active transport.

22. Describe the processes of endocytosis and exocytosis and indicate the differences between phagocytosis and pinocytosis.

23. Compare and contrast the following:
    a. passive and active transport
    b. facilitated and active transport.

Article 2.1: Bacteria and Other Organisms that Cause Infectious Diseases are Evolving Resistance to Drugs

Beginning in the late 1980's, an alarming increase in the incidence of tuberculosis (TB) has been documented worldwide. TB currently kills more than 2 million people each year. In the 30 or so years before the 1980s, the number of cases of TB has declined, at least in the developed world, largely as a result of treating TB with antibiotics, which are drugs intended to harm or kill bacteria and other microorganisms.

The evolution of drug-resistant strains in the bacterium that causes TB (*Mycobacterium tuberculosis*) is a disturbing trend. These strains are resistant to one or more antibiotics that traditionally were used to treat TB. Drug-resistant TB is deadly: As many as 80% of the people infected with multiple-resistant TB (MDR-TB) dies within two months of diagnosis—even with medical care. The problem with MDR-TB is particularly serious in five countries—Estonia, China, Latvia, Russia, and Iran—where 5% to 14% of all patients first diagnosis with TB are infected with multidrug-resistant strains.

Bacteria are continually evolving, even inside the bodies of human and animal hosts. Bacteria develop genetic resistance through mutations and through the acquisition of new genes from plasmids or viruses. When an antibiotic is used to treat a bacterial infection, a few bacteria may survive because they are genetically resistant to the antibiotic, and they pass these genes to future generation. As a result of selection (a change in the environment—the presence of the antibiotic), the bacterial population contains a larger percentage of antibiotic-resistant bacteria than before.

Poor prescribing practices by doctors and poor patient compliance with treatment are factors in the development of drug-resistant strains of the TB bacterium (as well as other bacteria). A person infected with TB must take three to ten pills of antibiotics each day for at least six months. After the first month of treatment, the person usually feels better; many patients decide to quit taking their medication at this point. When this happens, the TB bacteria is still lurking in their bodies and those with a resistance to the prescribed antibiotics rally. The evolution of a strain of bacteria resistant to several drugs is the worst-case scenario.

Article 2.2: Earth's Time Line

TIME (million years ago	GEOLOGICAL EVENT	BIOLOGICAL EVENT
4600	Creation of earth and our solar system	
4000	End of major impacts by other particles	
3800	Solidification of earth's crust, formation of first rocks	First unicellular organisms
3500	Condensation of water from atmosphere and formation of oceans	First photosynthetic organisms begin releasing oxygen, cynobacteria present
2400	Increase of oxygen levels leads to beginning of banded iron formations	
2000	Transition to stable aerobic hydrosphere and atmosphere	
1200		First eukaryotic organisms
1000		Beginning of multicellular eukaryotes
650		Mass extinction of stromatolites and many soft-bodies organisms
545		Cambrian explosion of hard-bodies organisms
515	Formation of Burgess Shale fossilization	
500		First vertebrates—fish
430		First land plant
420		First land animal—millipede
375	Formation of Appalachian Mountains	First sharks
350		Expansion of amphibians First insects First plants with roots—ferns
300		Expansion of reptiles Development of winged insects
250		Permian mass extinction
225		Bees, roaches, and termites evolve
200	Pangea begins to break apart	First crocodiles First mammals
145		Archaeopteryx
75		Rise of mammals
65	Cretaceous-Tertiary boundary	Mass extinction of dinosaurs
50		First monkeys
20		First apes, chimpanzees, and hominids
6		Split of ape and hominid lines

TIME (million years ago	GEOLOGICAL EVENT	BIOLOGICAL EVENT
4		*Australopithecus*, beginning of bipedalism
2		Widespread use of stone tools
1		Widespread use of fire
0.25	Most recent ice age	
0.2		First *Homo sapiens*
0.02		Cave paintings in Altamira Cave
0.01		First human permanent settlements
0.006		Writing developed in Sumeria

Figure 2.1: Geological Time as a 24-Hour Clock

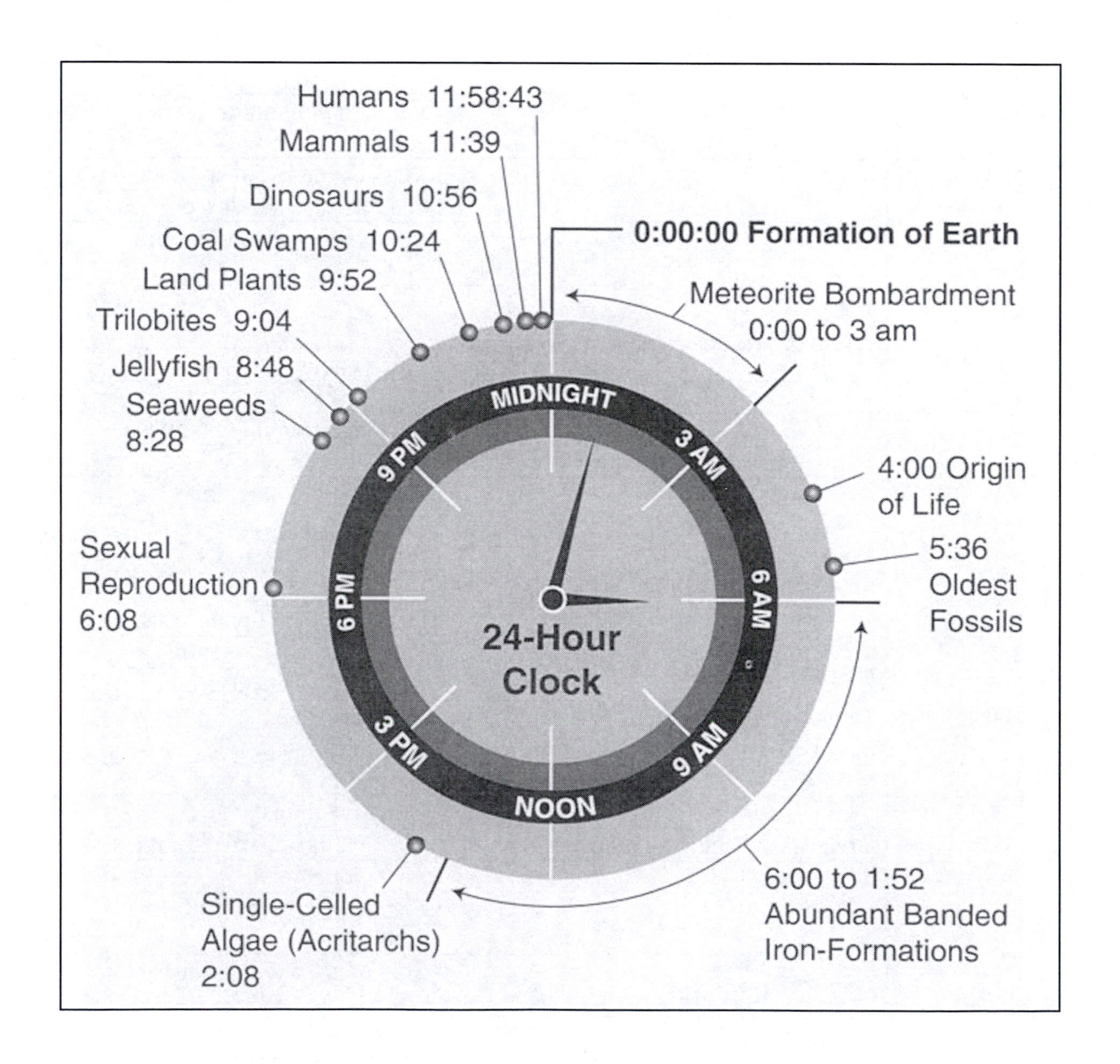

Worksheet 2.1: Prokaryotic Cells

Structure	Composition	Function	Other
Cell Wall			
Flagella			
Cell/Plasma Membrane			
Protoplasm			
Nucleoid (Region)			
Ribosome			

Worksheet 2.2: Eukaryotic Organelles

ORGANELLE	COMPOSITION	FUNCTION	OTHER
Cell Wall Extra-Cellular Matrix			
Cell Membrane Spot Desmosome Tight Junction Gap Junction Plasmodesmata			
Cytoplasm			
Nucleus			
Nucleolus			
Ribosome			

ORGANELLE	COMPOSITION	FUNCTION	OTHER
Endoplasmic Reticulum Rough (RER) Smooth (SER)			
Central Vacuole Contractile Vacuole Lysosome Vesicle Peroxisome			
Golgi Complex			
Chloroplast			
Mitochondria			

ORGANELLE	COMPOSITION	FUNCTION	OTHER
Cytoskeleton			
Microfilaments			
Intermediate Filaments			
Microtubules			
Flagella/Cilia			
Basal Bodies			
Centrioles			

Worksheet 2.3: Comparison of Cell Types

STRUCTURE	**PROKARYOTIC***	**EUKARYOTIC** Plant	**EUKARYOTIC** Animal
Cell Surface			
Cell Wall			
Plasma Membrane			
Organization of Genetic Material			
Genetic Material			
Nucleus			
Nucleolus			
Organelles			
Mitochondria			
Chloroplasts			
Ribosomes			
Endoplasmic Reticulum			
Golgi Complex			
Lysosomes			
Central Vacuole			
Other Vacuoles/Vesicles			
Centrioles			
Flagella/Cilia			

*Many structures are listed as absent in prokaryotic cells; however, their functions are often essential to the life of the cell. The plasma membrane generally performs the tasks of the eukaryotic organelles.

Figure 2.2: Comparison of Membrane Transport Mechanisms

Adapted from "Anatomy & Physiology, 9th Edition" by Seeley

Mechanism	Description	Substance Transported	Examples
Diffusion	Random movement of molecules (Brownian Motion) results in a net movement from areas of high to low concentration	Lipid soluble molecules dissolve in the bi-layer	O_2, CO_2, lipids such as steroids, and alcohol
Osmosis	A special type of diffusion where water moves through the selectively membrane	Water moves through the bi-layer	Water moves from the intestines to blood
Facilitated Diffusion	Requires proteins to move particles from areas of high to low concentrations TYPES		
	Channel Proteins: form tunnels that are specific to the particle crossing the membrane. The particles move from high to low concentrations	Substances that are too large, too polar or charged so they cannot dissolve in the bi-layer	Cl^- *Aquaporins*: Rapid diffusion of water; found in kidneys & plant root cells
	Gated (Ion) Channels: are opened/closed by chemical (ligand) or electrical (voltage) events thus allowing another particle to cross the membrane		Moves ions for nerve transmission
Active Transport	ATP combines with proteins specific to the particle and moves the particle from low to high concentrations	Substances that are accumulated in concentrations higher on 1 side of the membrane than another	Na^+, K^+, Ca^{+2}

Mechanism	Description	Substance Transported	Examples
Transporters	Requires proteins and may or may not require ATP TYPES		
	Uniports: Carrier proteins bind to the solute which causes the protein to change shape and the binded particle is moved across the membrane		Glucose, protons
	Symports: Carrier proteins move 2 particles (1 is traditionally an ion) in the same direction. Typically, the ion moves down its gradient thus supplying the energy to move the other particle against its gradient.		Na^+ and glucose
	Antiports: A specific carrier protein moves 2 particles in opposite directions. Both are moved from low to high concentrations thus requiring ATP.		Na^+ and K^+
Bulk Transport	Large molecules, particles of food and small cells are moved into and out of cells and requires ATP TYPES		
	Endocytosis: The plasma membrane forms a vesicle around the substances to be transported & the vesicle is taken into the cell	*Phagocytosis*: cells & solid particles *Receptor-Mediated*: specific substances *Pinocytosis*: molecules dissolved in water	White blood cells ingesting bacteria & cellular debris; amoeba eating Process used by most cells
	Exocytosis: Materials manufactured by the cell are packaged in secretory vesicles that fuse with the membrane & release the contents to the outside	Proteins & other water soluble molecules	Digestive enzymes, hormones, neurotransmitters, waste products

After Class Work

1. What characteristic of life fits the following descriptions?

    a. Organisms are composed of cells

    b. The leaves of a mimosa close when touched

    c. Molecules are built up

    d. Organisms pass on their organization to new generations

    e. Food is digested and waste products are excreted

    f. Organisms grow and are repaired if damaged

2. Label each of the following as an example of unity or diversity of life.

    a. Fungi absorb food and plants carry out photosynthesis.

    b. Homeostasis, metabolism and evolution are characteristics of living things.

    c. Life began with single cells.

    d. Living things consist of cells.

    e. Maple trees have broad, flat leaves and pine trees have leaves that resemble needles.

3. The following questions refer to either microfilaments or microtubules.

    a. Which is larger?

    b. Which is composed of tubulin protein?

    c. Which is found in centrioles, flagella and cilia?

    d. Which is important to the cytoskeleton?

4. How do these organelles work together?

a. lysosomes and food vacuoles

b. endoplasmic reticulum and Golgi complex

c. centrioles and cilia

d. ribosomes and endoplasmic reticulum

e. chloroplasts and mitochondria

5. State whether the following organelles are found only in prokaryotic cells, only in eukaryotic cells or in both.

a. Chromosome

b. True nucleus

c. Organelles

d. Plasma Membrane

e. Cell Wall

f. Ribosomes

g. Few structures

6. State whether the following pertains only to animal cells, only to plant cells or to both.

a. Centrioles

b. Plasma Membrane

c. Cell Wall

d. Large central vacuole

e. Small, numerous vacuoles

f. Mitochondria

g. Chloroplasts

h. Lysosomes

7. Select the key term that characterizes each of the following statements.

Key: diffusion, facilitated diffusion, osmosis, phagocytosis, active transport, pinocytosis

a. Requires ATP (cellular energy)

b. Passive process driven by kinetic energy of the molecules, no cellular energy expended and uses protein molecules

c. Movement of a substance from low concentration to high concentration

d. Engulf foreign substances

e. Movement of water through a semi-permeable membrane

f. Mechanism by which O_2 and CO_2 moves through the membrane

g. Provides for cellular uptake of solid or large particles into the cell

h. Provides for the engulfing and uptake of soluble substances into the cell

i. Mechanism that many small, lipid soluble substances use to move through the membrane

8. Label each of the following as describing active transport, diffusion or osmosis.

a. Algae in a pond became dehydrated.

b. A hypertonic solution draws water.

c. A red blood cells bursts in a person's bloodstream.

d. ATP is required by the cells.

e. Dye crystals spread out in a beaker of water.

f. Marine fish extrude salt through their gills.

g. Perfume is sensed from the other side of the room.

h. Plant root cells extract inorganic salt ions from the soil.

9. A semi-permeable sac containing 5% NaCl, 10% glucose and 15% albumin is suspended in a solution with the following composition: 11% NaCl, 12% glucose and 35% albumin. Assume the sac is permeable to all substances except the albumin. Using the choices below, indicate the correct answer.

Key: 1. moves into the sac 2. moves out of the sac 3. does not move

a. glucose

b. water

c. albumin

d. NaCl

10. Complete this chart to describe the effect of tonicity on red blood cells.

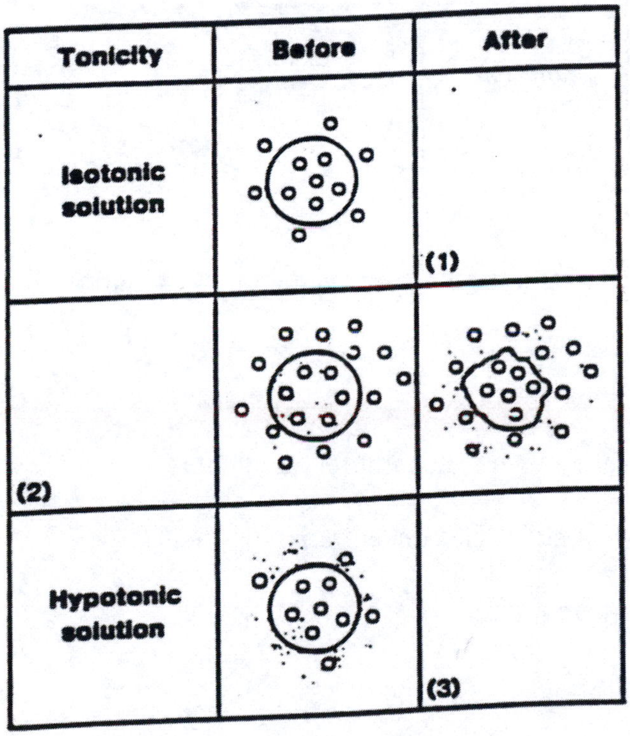

Tonicity	Before	After
Isotonic solution		(1)
(2)		
Hypotonic solution		(3)

11. Complete this chart to describe the effect of tonicity on plant cells.

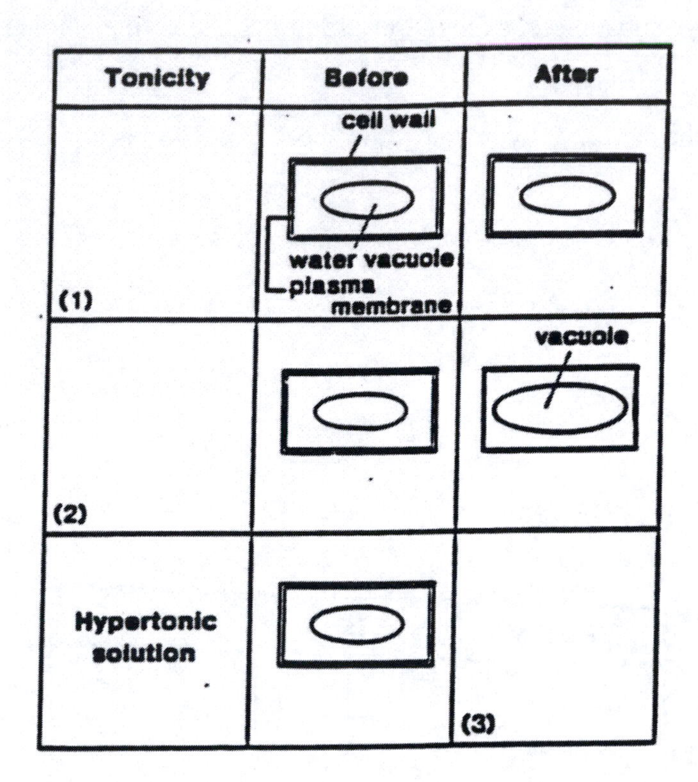

12. Label each of the following describing only facilitated transport only, active transport only or both.

a. Uses a carrier molecule

b. Substances travel down a concentration gradient

c. Substances move against the concentration gradient

d. Cellular energy is required

e. ATP is not required

13. Label each of the following as describing exocytosis or endocytosis.

a. Formation of vesicles by Golgi Apparatus

b. Materials leave the cell

c. Phagocytosis

d. Pinocytosis

Answers

1. a. organization
   b. react to stimuli
   c. metabolism
   d. reproduction
   e. metabolism
   f. metabolism
   g. metabolism

2. a. Diversity
   b. Unity
   c. Unity
   d. Unity
   e. Diversity

3. a. microtubule
   b. microtubule
   c. microtubule
   d. both

4. a. The food vacuole will fuse with the lysosome (digestive enzymes) so that intracellular digestion can occur.
   b. Products produced at the endoplasmic reticulum are sent to the Golgi complex for final packaging into a vacuole for cellular use or possible secretion.
   c. Centrioles organize the microtubules within cilia.
   d. Proteins are made at the ribosomes located on the endoplasmic reticulum. This complex is referred to as RER (rough endoplasmic reticulum).
   e. Carbohydrates made in chloroplasts are broken down in mitochondria.

5. a. both
   b. eukaryotic
   c. eukaryotic
   d. both
   e. both
   f. both
   g. prokaryotic

6. a. animal
   b. both
   c. plant
   d. plant
   e. animal
   f. both
   g. plant
   h. animal

7. a. phagocytosis, active transport, pinocytosis
   b. facilitated diffusion
   c. active transport
   d. phagocytosis
   e. osmosis
   f. diffusion
   g. phagocytosis
   h. pinocytosis
   i. diffusion

8. a. osmosis
   b. osmosis
   c. osmosis
   d. active transport
   e. diffusion
   f. active transport
   g. diffusion
   h. active transport

9. a. 1　b. 2　c. 3　d. 1

10. 1. same size and shape
    2. hypertonic solution
    3. cell is bursting

11. 1. isotonic
    2. hypotonic solution
    3. cell vacuole should be much smaller

12. a. both
    b. facilitated transport
    c. active transport
    d. active transport
    e. facilitated transport

13. a. exocytosis
    b. exocytosis
    c. endocytosis
    d. endocytosis

Sample Multiple Choice Questions

1. Pheromones, which are chemicals released by many animals in an effort to attract members of the opposite sex, are an example of:
    a. homeostatic mechanisms.
    b. a response to stimuli.
    c. a metabolic reaction.
    d. growth processes.
    e. reproduction.

2. The present system of biological classification contains three Domains. Which of the following does not belong to any of the domains?
    a. bacteria
    b. polio virus
    c. baker's yeast
    d. tulip
    e. elephant

3. A eukaryotic cell
    a. is usually smaller than a prokaryotic cell.
    b. has its DNA concentrated in one area of the cell without a nuclear membrane.
    c. typically has a cell wall in addition to a plasma membrane.
    d. is a bacteria-like organism.
    e. has a variety of membranous organelles.

4. Which of the following pairs is correctly matched?
    a. chloroplast: storage of enzymes
    b. lysosome: powerhouse of the cell
    c. nucleolus: site of ribosomal subunit synthesis
    d. Golgi complex: production of energy

5. A cellular structure found in plant but not animal cells is the
    a. chloroplast.
    b. ribosome.
    c. endoplasmic reticulum.
    d. cell membrane.
    e. mitochondria.

6. In a lipid bi-layer, _____ fatty acid tails face each other within the bilayer and form a regions that excludes water.
    a. hypertonic
    b. hypotonic
    c. hydrophilic
    d. hydrophobic

7. A patient who has had a severe hemorrhage accidentally receives a large transfusion of distilled water directly into a major blood vessel. You would expect this mistake to have
   a. no unfavorable effect as long as the water is free of bacteria.
   b. a serious, perhaps fatal consequences because there would be too much fluid to pump.
   c. a serious, perhaps fatal consequences because the red blood cells could shrink.
   d. a serious, perhaps fatal consequences because the red blood cells could swell and burst.
   e. no serious effect because the kidney could quickly eliminate any excess water.

8. The energy-requiring movement of materials against a concentration gradient is termed
   a. active transport.
   b. passive transport.
   c. facilitated diffusion.
   d. osmosis.
   e. all but a.

Answers to Multiple Choice Questions

1. B. The pheromones are attracting the male to the female, therefore the male is responding to a stimulus. Homeostasis refers to the maintenance of a constant, internal environment. Metabolism is the sum of all the chemical reactions that are occurring. Reproduction means to make more individuals/cells and the DNA must be passed on. Although the attraction will lead to reproduction, it is not.

2. B. A virus is not considered a living entity. The only thing that a virus is programmed to do is to make more viruses and to do this it must infect a host cell and use the metabolic processes of the host to replication. True bacteria are Eubacteria and bacteria that live in extreme environments are Archaea. All the others in the list are Eukarya.

3. E. Eukaryotic cells are characterized as large, complex with membrane bound organelles. Their DNA is organized into linear chromosomes contained in a nucleus. All the other choices explain a prokaryotic cell.

4. C. Chloroplast: photosynthesis. Lysosome: storage of digestive enzymes. Mitochondria: powerhouse of the cell or energy production. Golgi complex: modifies molecules and packages them into vesicles.

5. A. The chloroplast is what makes the plants green and is the site of photosynthesis. All cells must made proteins (ribosomes), transport nutrients in the cell (endoplasmic reticulum), transport materials into and out of the cell (cell membrane) and make useable energy, ATP (mitochondria).

6. D. Hydrophobic means to hate water. Each non-polar tail of the fatty acid must face each other—non-polar likes non-polar and polar likes polar. The head of the molecule is polar (hydrophilic) so they will face the watery environment or cytoplasm. Choices A and B refer to relative strengths of a solution. Hypotonic has less solutes in it than the hypertonic solution.

7. D. Blood contains various salts and sugars in it while distilled water has none. Therefore the blood is hypertonic and the water that is entering is hypotonic. Water (via osmosis, a form a passive transport) will enter the cells of the blood. Since an isotonic condition will never be reached, the cells will continue to swell and eventually burst. Answer C would be correct if a highly concentrated salt or sugar solution (the solution is hypertonic to the blood) were injected into the patient.

8. A. To move molecules against a gradient, a protein and cellular energy is required. This is the definition of active transport. All the other choices are passive transport where the molecules move with the gradient and Brownian motion supplies the energy of movement.

UNIT 3: ENERGY

Outline

TOPIC	SOLOMON	SACK.
I. From Previous Units		
A. Macromolecules		
1. Proteins		
B. Negative Feedback		
C. Oxidation and Reduction		
D. Energy Levels		
E. Membrane Structure		
F. Organelles		
a. Chloroplast		
b. Mitochondria		
II. Enzymes	156-162	
A. What are enzymes?		219-220
B. How do enzymes work?		222-223
1. Activation Energy		
2. Lock and Key Theory/Induced Fit		
C. Altering the Reaction Rate		
1. Substrate Concentration		
2. Enzyme Concentration		
3. Temperature		
4. pH		
5. Cellular Regulation of Pathways		223-226
(a) Enzymatic Competition		
(b) Inhibition Systems		
D. Coenzymes and Cofactors	155-156	221
1. Redox Reactions		
2. What are they?		
3. Important ones in metabolic pathways		
III. Physical Science		
A. Energy	148-153	
1. Definition		
2. Types		
3. Transformation		
a. Energy Laws		
b. Entropy		
B. Metabolism		
1. Catabolism (exergonic)		
2. Anabolism (endergonic)		

TOPIC	SOLOMON	SACK.
3. Coupled Reactions		
C. Measured by Heat		
D. ATP	153-154	
1. Structure		201-203
2. Cellular Use		
3. Phosphorylation of ADP	168	
a. Substrate Level		
b. Chemiosmotic	174-5, 194-5	
IV. Photosynthesis		261
A. Goals		
B. Sunlight	186	
1. Electromagnetic Spectrum		
a. Wavelength		
b. Photon		
2. Visible light		
C. Chloroplasts	85, 187-190	
1. Internal Structures		
2. Pigments		
D. Overview of Reactions	190-191	263-264
E. Light (Dependent) Reactions	191-196	265-271
1. Photosystems		
2. Non-cyclic Pathway		
3. Cyclic Pathway		
F. Fixation of Carbon (Light Independent)	196-200	
1. Calvin (C3) Cycle		272-277
a. Carbon dioxide fixation		
b. Reduction of CO_2		
c. RuBP Regeneration		
d. Photorespiration		
2. C4 Plants		278-279
3. CAM Plants		
V. Cellular Respiration		
A. Overview	166-167	
1. Breathing		
B. Glycolysis	168	248-252
C. Aerobic Respiration	82, 168-179	
1. Mitochondrial Structure		
2. Acetyl-CoA Formation (Transition Reaction)		252-255
3. Krebs Cycle		255-257
4. Electron Transport		257-259
5. Summary		244-248
6. Efficiency of System		

TOPIC	SOLOMON	SACK.
D. Anaerobic Processes	181-182	
1. Anaerobic Respiration		
2. Fermentations		
a. Yeast and Plants		
b. Bacteria and Animals		
c. Usefulness		
d. Efficiency of System		
E. Relationship to Photosynthesis		
F. Other Uses for Aerobic Respiration	183	
1. Metabolic Pool		
2. Weight Loss		
VI. Visual Supplements		
A. Tapes in Library		
1. Chemistry: The Cell and Energetics		
a. Part 7: ETS and ATP		
b. Part 11: ATP as an Energy Carrier		
c. Part 8: Photophosphorylation		
d. Part 9: C3 Cycle		
e. Part 10: C4 Cycle		
f. Part 5: Glycolysis		
g. Part 6: Oxidative Respiration		
2. Physiology		
a. Part 1: Lock & Key Model of Enzyme Action		
b. Part 5: Elec. Trans. & Oxid. Phosphorylation		
B. Thinkwell CD-ROM		
1. Inorganic and Organic Chemistry		
a. Enzymes		
b. Enzyme Action		
2. Photosynthesis		
a. Discovering Photosynthesis		
b. Adaptations for Photosynthesis		
c. The Light Reactions		
d. The Dark Reactions		
e. Photorespiration		
3. Respiration		
a. An Introduction to Respiration		
b. Glycolysis and Fermentation		
c. Aerobic Respiration		
d. ETC & Oxidative Phosphorylation		
VII. Exam		

Objectives

1. Describe the structure and function of enzymes and how enzymes lower the Energy of Activation.

2. Indicate the role of temperature, pH, concentrations and inhibitors play in altering enzyme activity.

3. Explain the function of coenzymes in metabolic pathways. Give examples of each

4. Define energy; list its two states and its basic forms, how it is measured, and how it is different from matter.

5. Compare and contrast the differences between potential and kinetic energy. Give examples of each.

6. State the first two laws of thermodynamics. Explain why life on this planet requires an outside source of energy and how energy is used to negate entropy.

7. Describe the major characteristics of the various types of metabolic reactions: anabolism vs. catabolism and endergonic vs. exergonic.

8. Describe the structure and function of ATP. Explain how ATP holds energy for the cell while other macromolecules store the energy for the cell.

9. Contrast the amounts of energy stored in glucose, NADH or $FADH_2$ and ATP.

10. Explain the processes of substrate-level phosphorylation and chemiosmosis.

11. State the generalized reaction for photosynthesis and indicate the accomplishments of photosynthesis which are essential to life on this planet. Be able to identify which reactants are oxidized and reduced.

12. Describe the entire electromagnetic spectrum and indicate their relative energy levels.

13. Describe the structures and functions of the components of a chloroplast. Define the role of the pigments and how they "operate" in a Photosystem.

14. Indicate the locations, reactants and products of the light (dependent) and light independent (carbon fixation) reactions of photosynthesis. Describe how the light (dependent) reactions drive the light independent (carbon fixation) reactions.

15. Describe the route taken by an electron in the cyclic and non-cyclic pathways of the light dependent reactions of photosynthesis. Indicate the reactants and products of the pathways.

16. For the Calvin Cycle: state its purpose, reactants and products. Describe the three phases of the cycle. Explain the role of ATP and NADPH.

17. Define photorespiration and explain why this process is detrimental to C3 plants.

18. Contrast C4 and CAM plants to C3 plants. Give examples of each.

19. State the generalized reaction for aerobic respiration and its function.

20. Compare and contrast rapid and slow oxidation

21. Distinguish between aerobic and anaerobic.

22. State the generalized reactions for aerobic respiration. Identify which reactants are oxidized and reduced. Explain the total energy yield from 1 glucose molecule.

23. List the major reactants, products, location and purpose of glycolysis.

24. Describe the transition reaction. Indicate its location.

25. List the major reactants, products, location and purpose of the Krebs Cycle.

26. Explain the two major accomplishments of the mitochondrial electron transport system. Indicate its location.

27. Define anaerobic respiration.

28. Explain fermentation in terms of its general usefulness, products and energy yield.

29. List the three processes used by different cells to oxidize NADH after glycolysis is completed. State whether the processes are examples of aerobic respiration or fermentation.

30. Compare the efficiency of aerobic respiration to fermentation.

31. Explain how aerobic respiration can be used in both the catabolism and anabolism of macromolecules.

32. Explain the interconnection between energy and matter when discussing photosynthesis and aerobic respiration.

Article 3.1: Generating ATP

Virtually every cell in every organism relies on energy from ATP molecules. Every movement we make, every thought or memory we have, and every molecules our cell manufacture depend directly on ATP energy. Cells generate ATP by phosphorylation, that is, by adding a phosphate group to ADP. A cell has two ways to do this: chemiosmotic phosphorylation, also called chemiosmosis, and substrate-level phosphorylation.

In 1978, British biochemist, Peter Mitchell, was awarded the Nobel Prize for developing the theory of chemiosmosis. Mitchell's theory describes how cells use the potential energy in concentration gradients to make ATP. A concentration gradient of a solute stores energy due to the tendency of the solute molecules to diffuse from where they are more concentrated to where they are less concentrated. The theory of chemiosmosis centers on membranes, and in particular on the activity of ATP syntheses, protein complexes (clusters) that reside in membranes. ATP synthases synthesize ATP using the energy stored in concentration gradients of H^+ ions (that is, protons) across membranes. Cells generate most of their ATP in this way.

Let's look at an overview of the relationship between membrane structure and chemiosmotic ATP synthesis. ATP synthase is built into the same membrane as the molecules of an electron transport chain. This structural connection allows the energy that hydrogen's electrons deliver to the electron transport chain to drive the production by ATP by the ATP synthase. Within the chain, redox reactions release energy from electrons cascading down the series of electron carriers. As these exergonic reactions release energy, some of the proteins built into the chain use the energy to actively transport H^+ ions across the membrane. This flow results in an increased concentration gradient of H^+ ions across the membrane. The ATP synthase then uses the potential energy of the concentration gradient to drive the endergonic reaction that generative ATP from ADP and phosphate.

Substrate-level phosphorylation is much simpler than chemiosmosis and does not involve a membrane. In substrate-level phosphorylation, an enzyme transfers a phosphate group from an organic molecules to ADP. The substrate is one of several substances produced as cellular respiration converts glucose to CO_2. The reaction occurs because the bond holding the phosphate group in the substrate molecule is less stable than the new bond holding it to ATP. The reaction produces a new organic molecule and a molecule of ATP. Substrate-level phosphorylation accounts for only a small percentage of the ATP that a cell generates.

Worksheet 3.1: Connections Between the Light Dependent and Light Independent Reactions:

NAME:	NAME:

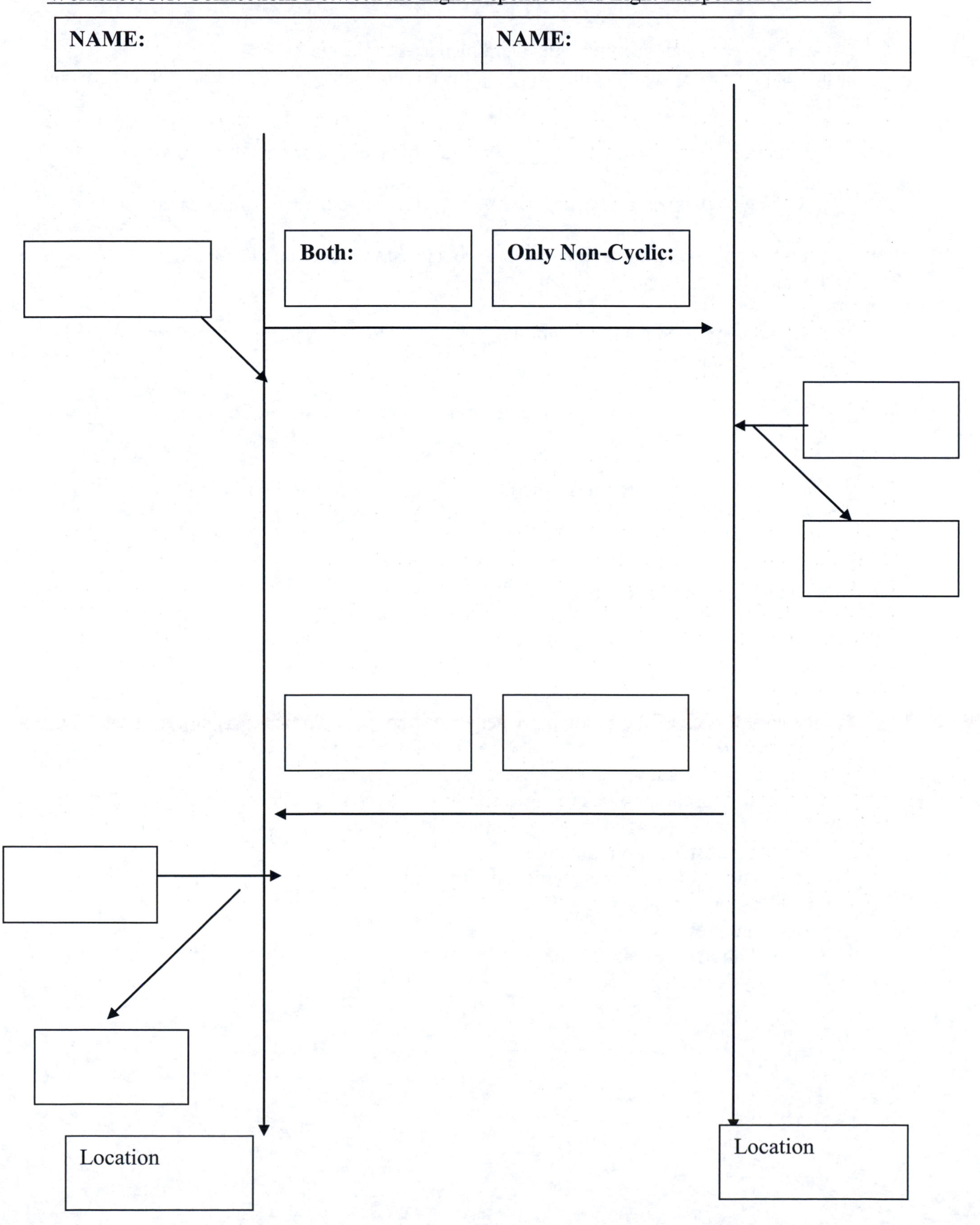

After Class Work

1. Complete each statement with the word increases or decreases.
   a. The presence of an enzyme ________ the required energy of activation for a chemical reaction.

   b. Generally more substrate ________ the rate of an enzymatic reaction.

   c. Raising the temperature over 50° C ________ the rate of an enzymatic reaction.

   d. Lowering the pH for an enzyme that works best in highly acidic conditions ________ the rate of an enzymatic reaction.

   e. Introducing a competitive inhibitor ________ the availability of an enzyme for its normal substrate.

   f. By feedback inhibition, the release of a product made by an enzyme ________ the activity of the enzyme.

2. Match each substance to one of the following terms: cofactor, FAD, NAD
   a. can accept only a H^+

   b. can accept 2 hydrogen ions

   c. Mg^{+2}

3. Indicate for each of the following whether it is an example or description of anabolism or catabolism.
   a. Chemical reaction which releases energy
   b. Chemical reactions which requires energy
   c. Break down of polypeptides into amino acids
   d. Build up of starch from glucose
   e. Production of fats from glycerol and fatty acids
   f. Oxidation reaction
   g. Reduction reaction
   h. Dehydration synthesis reaction
   i. Hydrolysis reaction

4. Consider the following 2 reactions, where A, B, C, D, E and F refer to different cellular molecules.

   Reaction 1: A → B + C + energy

   Reaction 2: D + E + energy → F

   a. Which reaction is synthetic? degradative?

   b. How are reactions 1 and 2 related?

   c. How can NADH link the 2 reactions?

   d. How are the 2 reactions in a cell dependent on an outside source of matter and energy?

5. Are the following reactions involving coenzymes oxidation or reduction reactions?

   a. NAD + H → NADH

   b. $FADH_2$ → FAD + 2H

6. For the following reactions, which releases energy and which stores energy?

   a. ADP + P → ATP

   b. ATP → ADP + P

7. Label each of the following events as being part of the light dependent reaction or the light independent reaction.

   a. ATP is produced.

   b. Calvin Cycle takes place.

   c. Chlorophyll molecules are excited.

   d. Fixation of carbon dioxide occurs.

   e. Has cyclic and noncyclic pathways.

   f. NADPH is formed.

   g. Occurs in Photosystems I and II.

   h. PGAL is a product.

8. State whether each of the following substances or events is found in the thylakoids or stroma.

   a. CO_2

   b. H_2O

   c. ATP produced

   d. ATP used up

   e. NADPH produced

   f. NADPH is a reactant

   g. RuBp

   h. PGAL

   i. oxygen

   j. cytochrome system

   k. enzymes

The following questions refer to glycolysis.

9. Is oxygen required?
10. Where in the cell does glycolysis occur?
11. Starting with 1 molecule of glucose, how many pyruvate(s) is/are produced?
12. What is the other name for pyruvate?
13. During this process, what happens to the NAD molecules?
14. How many ATP molecules are "used"?
15. How many ATP molecules are produced?
16. What are the final products of glycolysis?

The following questions refer to fermentation reactions.

17. Is oxygen required?
18. Where in the cell is fermentation taking place?
19. What is/are the end product(s) of fermentation in most animal cells?
20. What is the usefulness of fermentation in animals?
21. What does a build up of the molecule(s) formed in #19 feel like in your muscles?
22. What is/are the end product(s) of fermentation in yeast?
23. What is the usefulness of fermentation in yeast?
24. Where did the NADH molecules that are used come from?
25. Where do the NAD molecules go?
26. Starting with 1 glucose, what is the net yield of ATP?

The following questions refer to the generation of Acetyl-CoA (transition reaction).

27. Where in the cell are the reactions taking place?
28. Is oxygen required?
29. What is the role of Coenzyme A (CoA)?
30. Why do we say "starting with 2 pyruvates"?
31. What are the end products of this short pathway?

The following questions refer to the Krebs Cycle.

32. Where in the cell is the cycle taking place?
33. Is oxygen required?
34. How does the acetyl-CoA enter the cycle? What is the alternate name for this pathway?
35. Why do we say "starting with 2 acetyl-CoAs"?
36. What are the end products of the Kreb's Cycle?

The following questions refer to the Electron Transport System.

37. Where in the cell is the pathway taking place?
38. What are the molecules that are used in electron transport? Where did they come from?
39. How do the molecules mentioned in #38 produce ATP?
40. For each molecule of NADH, how many molecules of ATP are produced?
41. For each molecule of $FADH_2$, how many molecules of ATP are produced?
42. What is the role of oxygen?
43. What forms as a direct result of oxygen?

The following questions refer to the complete pathways.

44. Starting with 1 molecule of glucose, how many ATPs are produced by anaerobic respiration?
45. Starting with 1 molecule of glucose, how many ATPs produced by aerobic respiration?
46. When a typical fat molecule is digested approximately 30, 2-carbon fragments are produced. Each fragment can be converted to acetyl-CoA and processed through the Kreb's Cycle. Assuming that electron transport will also occur, how many ATPs can be produced from the energy released by the fat?

47. Explain why eating a lot of carbohydrates could make a person fat.

48. To contrast photosynthesis with cellular respiration, complete this chart.

DESCRIPTION	PHOTOSYNTHESIS	CELLULAR RESPIRATION
Cell Type		
Organelle		
Reaction Type		
CO_2		
Oxygen		
Glucose		

49. Label each of the following as describing the Transition Reaction, Kreb's Cycle or the Electron Transport System.

a. Acetyl Group and CoA are combined.

b. Associated with the process of oxidative phosphorylation.

c. Begins with a molecule of citrate.

d. Connects glycolysis to the Kreb's Cycle.

e. Located in the cristae of the mitochondrion.

f. Located in the matrix of the mitochondrion.

g. Oxidation produces NADH and $FADH_2$ molecules.

h. Oxygen in the final acceptor.

i. Pyruvate is oxidized and converted to an acetyl group.

Answers

1. a. decreases
   b. increases
   c. decreases
   d. increases
   e. decreases
   f. decreases

2. a. NAD
   b. FAD
   c. cofactor

3. a. catabolism
   b. anabolism
   c. catabolism
   d. anabolism
   e. anabolism
   f. catabolism
   g. anabolism
   h. anabolism
   i. catabolism

4. a. synthetic: reaction 2; degradative: reaction 1
   b. The energy released from reaction 1 can drive reaction 2.
   c. It can carry H+ released from reaction 1 to make the ATP used for the energy required in reaction 2.
   d. Matter offers the raw material for synthesis; energy is needed for this building.

5. a. reduction
   b. oxidation

6. a. stores energy
   b. releases energy

7. a. light dependent
   b. light independent
   c. light dependent
   d. light independent
   e. light dependent
   f. light dependent
   g. light dependent
   h. light independent

8. a. stroma
   b. thylakoids
   c. thylakoids
   d. stroma
   e. thylakoids
   f. stroma
   g. stroma
   h. stroma
   i. thylakoids
   j. thylakoids
   k. stroma

9. no
10. cystol
11. 2
12. pyruvic acid
13. become NADH
14. 2
15. 4
16. 2 pyruvates, NADH, 2 ATPs

17. no
18. cytosol
19. lactic acid
20. quick/emergency energy
21. achy, tired
22. CO_2 and ethanol
23. bread rise, wine, beer, distilled spirits
24. glycolysis
25. back to glycolysis so that ATP can be generated
26. 2
27. mitochondria, matrix
28. yes
29. join to an acetyl group and bring it into Kreb's Cycle
30. each glucose becomes 2 pyruvates & each must complete the process
31. CO_2, NADH, acetyl-CoA
32. mitochondria, matrix
33. yes
34. acetyl (carried in by CoA) joins to oxaloacetate and becomes citric acid, thus the other name for the Kreb's Cycle comes from the first product that is formed
35. 2 were formed from the 2 pyruvates of glycolysis
36. CO_2, NADH, $FADH_2$, ATP
37. mitochondria, cristae
38. NADH, $FADH_2$; previous steps
39. deposits the H's onto cytochromes that are an ETS
40. 3
41. 2
42. final electron and H^+ acceptor
43. forms water
44. 2
45. 36 for most cells (38 for hear & liver)

46. Each 2 carbon fragment when processed through the Kreb's Cycle results in the formation of 1 ATP, 3 NADH and 1 $FADH_2$. As a result of electron transport, the 3 NADH's can be "cashed in" for 9 ATPs (remember each NADH is worth 3 ATP's) and the 1 $FADH_2$ for 2 ATPs. In all, each 2 carbon fragment is "worth" 12 ATPs; so, 30 fragments release enough energy to produce 360 ATPs. So fats provide a lot of energy, and of course also lots of calories.

47. Carbohydrates are broken down to acetyl-CoA units that can be joined to give fatty acids.

48.

DESCRIPTION	PHOTOSYNTHESIS	CELLULAR RESPIRATION
Cell Type	plant	plant and animal
Organelle	chloroplast	mitochondria
Reaction Type	reduction	oxidation
CO_2	reactant	product
Oxygen	product	reactant
Glucose	product	reactant

49. a. Transition Reaction
   b. Electron Transport System
   c. Kreb's Cycle
   d. Transition Reaction
   e. Electron Transport System
   f. Kreb's Cycle
   g. Kreb's Cycle
   h. Electron Transport System
   i. Transition Reaction

Sample Multiple Choice Questions

1. Which of the following accurately represents the relationship between the terms anabolism, catabolism, and metabolism
    a. anabolism = catabolism
    b. metabolism = catabolism
    c. catabolism = anabolism + metabolism
    d. anabolism + catabolism + metabolism
    e. metabolism = catabolism + anabolism

2. Enzymes are important biological catalysts because they _____ a biochemical reaction.
    a. supply the energy to initiate
    b. increase the free energy of
    c. lower the entropy of
    d. decrease the enthalpy of
    e. lower the activation of

3. If one continues to increase the temperature in an enzymatic reaction, the rate of the reaction
    a. does not change.
    b. increases and then levels off.
    c. decreases and then levels off.
    d. increases and then decreases rapidly.
    e. decreases and then increases rapidly.

4. During chemiosmosis, _____ are transferred from NADH and $FADH_2$ to electron acceptor molecules, and the energy released is used to create a(n) _____ gradient across the inner mitochondrial membrane.
    a. protons, electrons
    b. electrons, protons
    c. ATP molecules, ADP molecules
    d. ADP molecules, ATP molecules
    e. water molecules, oxygen

5. During the reactions of photosynthesis, _____ is reduced and _____ is oxidized.
    a. O_2, $C_6H_{12}O_6$
    b. CO_2, $C_6H_{12}O_6$
    c. H_2O, $C_6H_{12}O_6$
    d. CO_2, H_2O

6. During the light dependent reactions, a constant supply of electrons is provided by
   a. water.
   b. oxygen.
   c. the sun.
   d. chlorophyll.
   e. glucose.

7. The first step in the Calvin Cycle is the attachment of carbon dioxide to
   a. rubisco.
   b. PGAL.
   c. phosphoglycerate.
   d. RuBP
   e. glucose.

8. Glycolysis yields a net energy profit of _____ ATP molecules per molecule of glucose.
   a. 0
   b. 1
   c. 2
   d. 4
   e. 36

9. The role of oxygen molecules required for aerobic respiration is to
   a. accept the low energy electrons at the end of ETS.
   b. form ATP,
   c. produce CO_2.
   d. store high energy electrons to pass through complexes of ETS.
   e. accept electrons directly from NADH and $FADH_2$.

10. The production of alcohol or lactate from pyruvate during _____ occurs as a means of regenerating _____ from _____.
    a. aerobic respiration, NAD, NADH
    b. fermentation, NAD, NADH
    c. fermentation, NADH, NAD
    d. fermentation, ADP, ATP
    e. aerobic respiration, ATP, ADP.

Answers to Multiple Choice Questions

1. E. Metabolism is the sum of all the chemical reactions occurring in a cell. Metabolism can be broken down into two forms. Catabolism is the breaking down of large molecules into smaller ones (releases energy) and anabolism builds larger molecules from smaller ones (stores energy).

2. E. Enzymes are reaction helpers that lower the amount of energy that must be put into the system to make the reaction happen. This energy is called the energy of activation. Everything else about the reaction stays the same: the feasibility of the reaction, the amount of energy in the reactants and products.

3. D. All reactions are based on effective, random collisions. With an enzymatic reaction, the enzyme and substrate must still meet and with an increase in temperature, the movement of the molecules will also increase, thus the rate of the reactions will increase. However, after a certain point, the temperature is too high the protein enzyme will be denatured. Since the enzyme no longer has the correct shape, it can no longer do its job. Even though the increased temperature has greatly increased the movement of the enzyme and substrates, the reaction will now stop.

4. B. The role of chemiosmosis is to make ATP from its subunits, ADP and P_i. This is accomplished by high-energy electrons moving through an electron transport system (ETS) and releasing their energy. This released energy is then used to pump protons (H+) across the membrane to establish a gradient. The H+s will want to establish equilibrium and in the process of repassing through the membrane will activate the ATP synthase.

5. D. Reduction is the addition of hydrogens. CO_s will accept hydrogens and become glucose. Oxidation is the loss of hydrogen. Water will lose its hydrogens (which are accepted by the CO_2). Water, then, created the product, oxygen.

6. A. The purpose of the light dependent reactions is to create ATP and NADPH. The source of the hydrogen (which will be split into a proton and electron) is water. The electrons will be send to a high energy state by the Photosystems and by moving through various transport systems, the products will be made.

7. D. The CO_2 is initially fixed to RuBP by the enzyme rubisco (choice A). The true end product of the cycle is choice B, PGAL, which will be converted into choice E, glucose.

8. Glycolysis created a total (gross) of 4 ATP molecules. However, 2 ATPs are required to activate the reaction, so the yield (net) is 2. The complete aerobic breakdown of glucose will produce 36 ATPs.

9. Oxygen is the final electron acceptor in the electron transport system of aerobic respiration. The source of the electrons is either NADH or $FADH_2$. When oxygen accepts the electrons (and H+) from ETS, water is produced.

10. These products are produced by the process called fermentation. The purpose of any fermentation process is to regenerate NAD (from NADH) so the glycolysis (and the production of small amounts of ATP) can occur.

UNIT 4: DNA & CHROMOSOMES

Outline

TOPIC	SOLOMON	SACK.
I. From Previous Units		
A. Bonds		
1. Covalent		
2. Hydrogen		
B. Enzyme Specificity		
C. Asexual and Sexual Reproduction		
D. Organelles		
1. Nucleus		
2. Nucleolus		
3. Centrioles		
4. Mitochondria		
5. Ribosomes		
II. Deoxyribonucleic Acid	252-258	
A. Molecular Criteria for a Gene		
B. Search for the Answer		
1. A New Molecule		
(a) Friedrich Miescher		
(i) Nucleotides	61-67	
(ii) Tetronucleotide Hypothesis		
2. Genes Code for Proteins		
(a) Archibald Garrod		
(b) G. W. Beadle & E. L. Tatum		
3. Transformation Properties		
(a) Frederick Griffith		
(b) Oswald Avery		
(c) Alfred Hershey & Margaret Chase		
4. Complementary Base Pairing		202-206
(a) Erwin Chargaff		
5. Size and Structure		
(a) Maurice Wilkins & Rosalind Franklin		
6. Putting it all Together		
(a) James Watson & Francis Crick		
(i) Double Helix		
(ii) Bonds		
C. Replication	259-267	
1. The Process		208-209
2. Semi-Conservative Nature		
3. Accuracy		

TOPIC	SOLOMON	SACK.
4. Telomeres		
III. Gene Expression		
A. Activity of Genes	271-273	
1. Genes to Enzymes		
2. One Gene to One Enzyme		
3. One Gene to One Polypeptide		
4. "Modern" Definition		
B. An Intermediate Molecule		
1. Ribonucleic Acid	273-275	
a. Structure		
b. Types		
C. Central Dogma		
D. The Genetic Code	276-277	
1. Triplet Codons		
2. Non-overlapping		
3. Exact Meaning		
4. Degenerative/Redundant		
5. Punctuation		
6. Universality		
E. Transcription	277-282	
1. The Process		209-212
2. Processing Eukaryotic mRNA		
a. Introns and Exons		
b. Caps and Tails		
F. Translation	282-287	
1. Additional RNAs Involved		
a. rRNA		
b. tRNA		
2. The Process		213-214
a. Initiation		
b. Elongation		
c. Termination		
G. Errors in the DNA	288-290	
1. Gene Mutation		215-217
a. Frameshift		
b. Point		
c. Positional Changes		
2. Results of Mutation		
a. Silent		
b. Neutral		
c. Missense		
d. Nonsense		
3. Germ Line vs. Somatic Cells		

TOPIC	SOLOMON	SACK.
4. Causes		
5. Protection		
IV. Mitosis		
A. Importance of Cell Divisions		
B. Prokaryotic Cells	213-214	
1. Chromosome Structure		
2. Binary Fission		
C. Eukaryotic Cells		
1. Chromosome Structure	205-207	
2. Overview		
3. Cell Cycle	203	
a. Interphase		
(i). G_1		
(ii). S		
(iii). G_2		
b. M Phase		
(i). Mitosis		
(ii). Cytokinesis		
4. Mitosis in Animal Cells	209-213	
(a) Prophase		
(b) Metaphase		
(c) Anaphase		
(d) Telophase		
(i) Cytokinesis		
5. Mitosis in Plant Cells	213	
D. When Mitosis Goes Wrong	215-216	
V. Meiosis		
A. Overview	217	
1. Importance		
2. Homologous Chromosomes	218	
(a) Diploid and Haploid Numbers		
3. Animal Life Cycle		
B. The Stages	218-222	
1. Meiosis I		
2. Meiosis II		
C. Comparison of Meiosis and Mitosis		
D. Meiosis in Humans	223	
1. Spermatogenesis		
2. Oogenesis		
(a) Mitochondrial Inheritance		
E. Sources of Variation		
1. Crossing Over		
2. Independent Assortment		

TOPIC	SOLOMON	SACK.
3. Fertilization		
4. Mutations		
VI. When Meiosis Goes Wrong		
A. Diagnosis	337, 352-354	
1. Amniocentesis		
2. Chorionic Villus Sampling		
3. Karyotype		
B. Normal Human Chromosome Make-up	239-240	
C. Wrong Chromosome Number	340-346	
1. Non-disjuntion		
(a) Autosomal Disorders		
(i) Downs Syndrome		
(b) Sex Chromosome Disorders		
(i) Turners		
(ii) Trisomy X		
(iii) Klinefelters		
(iv) XYY Males		
(v) Role of the Y Chromosome		
D. Structural Problems		
1. Translocation		
2. Deletions		
(a) Cri-du-chat		
3. Additions		
(a) Fragile X		
VII. Visual Supplements		
A. Tapes in Library		
1. National Geographic: Discovering the Cell		
2. Cell Division, Heredity, Genetics & Reproduction		
(a) Part 15: DNA Replication		
(b) Part 16: Transcription of a Gene		
(c) Part 17: Protein synthesis		
(d) Part 12: Mitosis		
(e) Part 13: Meiosis		
(f) Part 14: Crossing Over		
(g) Part 19: Spermatogenesis		
(h) Part 20: Oogenesis		
3. 624: The Genetic Code		
B. Models		
1. DNA for Student Use		
2. DNA Made Easy (DNA to RNA to Proteins)		

TOPIC	SOLOMON	SACK.

C. Thinkwell CD-ROM
  1. Inorganic & Organic Chemistry
    a. (Lipids) and Nucleic Acids
  2. Molecular Genetics
    a. Discovering DNA
    b. DNA Structure Revealed
    c. Events of DNA Replication
    d. Transcription
    e. Translation
    f. Protein Synthesis Overview
  3. Cell Reproduction
    a. An Introduction to Cell Cycle & Mitosis
    b. Meiosis
    c. Understanding Meiosis

VIII. Exam

Objectives

1. Discuss the location, chemical composition, structure and function of DNA. Be able to utilize the following terms: primers, 1', 3', 5', nucleotide, adenine, guanine, cytosine, thymine, purine, pyrimidine, sugar-phosphate backbone, covalent and hydrogen bonds.

2. Explain the contribution of James Watson and Francis Crick to the development of the structure of the DNA molecule.

3. Describe the process and results of DNA replication. Include the roles of helicase and other proteins in producing the replication fork. Discuss the roles of primase, DNA polymerase, topoisomerase and DNA ligase and how they function on the continuous (leading) strand and the discontinuous (lagging) strand.

4. Define semi-conservative replication and explain how it ensure the integrity of the genetic code.

5. Indicate the similarities and differences between DNA and RNA.

6. Explain how information is stored in DNA and the relationship between the nucleic acid sequence in DNA and the amino acid sequence of a protein.

7. Explain how the message in the DNA is transcribed and how mRNA is processed before it leaves the eukaryotic nucleus. Be able to explain the role of promoter and terminator regions, exons and introns, upstream and downstream sequences and start and stop codons.

8. Explain how mRNA is translated into the primary structure of a protein.

9. Given a DNA sequence and a genetic dictionary, indicate the following sequences: mRNA, tRNA and amino acids.

10. Define mutation. Be able to discuss various types of mutations, causative agents and how DNA is protected from mutations.

11. Indicate the functions of cell division in single celled and multicellular organisms and why being multicellular is an advantage. (Relate to genetic diversity)

12. Indicate how cell division occurs in prokaryotic cells and what it is called.

13. Define the following terms in relation to eucaryotic cells: chromatin, chromosomes, sister chromatids, homologous chromosomes, centromere, centrioles, polar spindle fibers, synaptonemal complex, and chiasmata.

14. Differentiate between diploid and haploid numbers.

15. List and explain the major phases and events of the cell cycle.

16. Explain the purpose of mitosis.

17. List and explain the major phases and events of mitosis in plant and animal cells.

18. Describe the life cycle of animals.

19. Explain the purpose of meiosis.

20. List and explain the major phases and events of meiosis.

21. Describe how the processes of crossing over and independent assortment result in genetic diversity.

22. Explain why some genes are more likely to be inherited together. (Relate to crossing over and independent assortment)

23. Compare and contrast oogenesis and spermatogenesis.

24. Compare and contrast the outcomes of mitosis and meiosis. Explain how meiotic events contribute to genetic diversity.

25. Describe how a karyotype is prepared and how it is used.

26. Indicate the chromosomal similarities and differences between human males and females. Indicate the symptoms and cause of the following syndromes: Turners, Trisomy X, Klinefelter, XYY, Downs, Cri-du-Chat and Fragile X.

27. Describe the role of the Y chromosome in producing a male looking individual.

Article 4.1: Matrix of Biology & Computer Science Concepts

College Science Teacher; November 2002; page 177

BIOLOGY	COMPUTER SCIENCE
1. Digital alphabet consists of bases A, C, T, G	1. Digital alphabet consists of 0, 1
2. Codons consist of three bases	2. Computer bits form bytes
3. Genes consist of codons	3. Files consist of bytes
4. Promoters indicate gene locations	4. File-allocation table indicates file locations
5. DNA information is transcribed into pre-mRNA and processed into mRNA	5. Disc information is translated onto a screen or paper.
6. mRNA information is translated into proteins	6. RAM information is translated onto a screen or paper
7. Genes may be organized into operons or groups with similar promoters	7. Files are organized into folders
8. "Old" genes are not destroyed; their promoters become non-functional	8. "Old" files are not destroyed; references to their location are deleted
9. Entire chromosomes are replicated	9. Entire discs can be copied
10. Genes can diversify into a family of genes through duplication	10. Files can be modified into a family of related files
11. DNA from a donor can be inserted into host chromosomes	11. Digital information can be inserted into files
12. Biological viruses disrupt genetic instructions	12. Computer viruses disrupt software instructions
13. Natural selection modified the genetic basis of organism design	13. Natural selection procedures modify the software that specifies a machine design
14. A successful genotype in a natural population out competes others	14. A successful web site attracts more "hits" than others

Figure 4.1: The Genetic Dictionary

Table of mRNA codons					
First Base	Second Base				Third Base
	U	C	A	G	
U	phenylalanine	serine	tyrosine	cysteine	U
	phenylalanine	serine	tyrosine	cysteine	C
	leucine	serine	STOP	STOP	A
	leucine	serine	STOP	tryptophan	G
C	leucine	proline	histidine	arginine	U
	leucine	proline	histidine	arginine	C
	leucine	proline	glutamine	arginine	A
	leucine	proline	glutamine	arginine	G
A	isoleucine	threonine	asparagine	serine	U
	isoleucine	threonine	asparagine	serine	C
	isoleucine	threonine	lysine	arginine	A
	START methionine	threonine	lysine	arginine	G
G	valine	alanine	aspartate	glycine	U
	valine	alanine	aspartate	glycine	C
	valine	alanine	glutamate	glycine	A
	valine	alanine	glutamate	glycine	G

Worksheet 4.1: Protein Synthesis

A. Using the following DNA sequence, show me the steps that lead to the mRNA and protein (amino acid sequence).

coding strand 5'- GCTCAATGCCCGGATACCATGCGAAAGGGAGCTCGATCTAATGCA- 3'
template strand 3'- CGAGTTACGGGCCTATGGTACGCTTTCCCTCGAGCTAGATTACGT- 5'

mRNA:

amino acids: (Remember start and stop codon)

B. Suppose a point mutation occurred that changed a cytosine 'C' to an adenine 'A'. What would the resulting protein amino acid sequence be?

coding strand 5'- GCTCAATGCCCGGATAACATGCGAAAGGGAGCTCGATCTAATGCA- 3'

template strand 3'- CGAGTTACGGGCCTATT GTACGCTT TCCCTCGAGCTAGATTACGT- 5

mRNA:

amino acids:

Figure 4.2: Cell Numbers

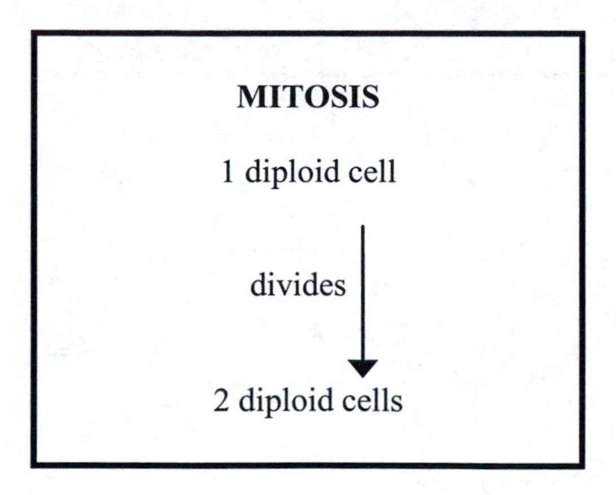

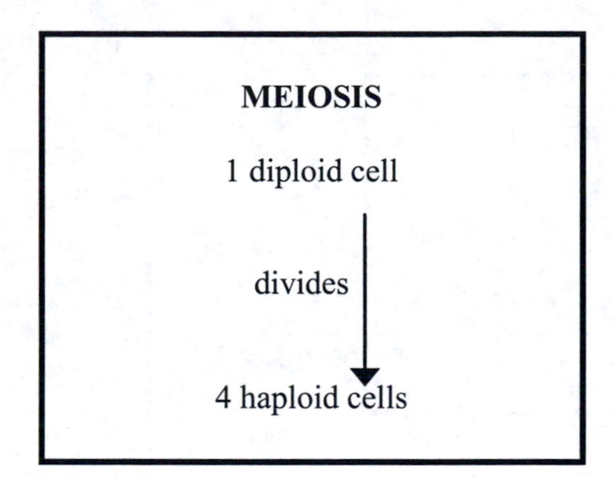

Figure 4.3: Chromatids and Chromosomes

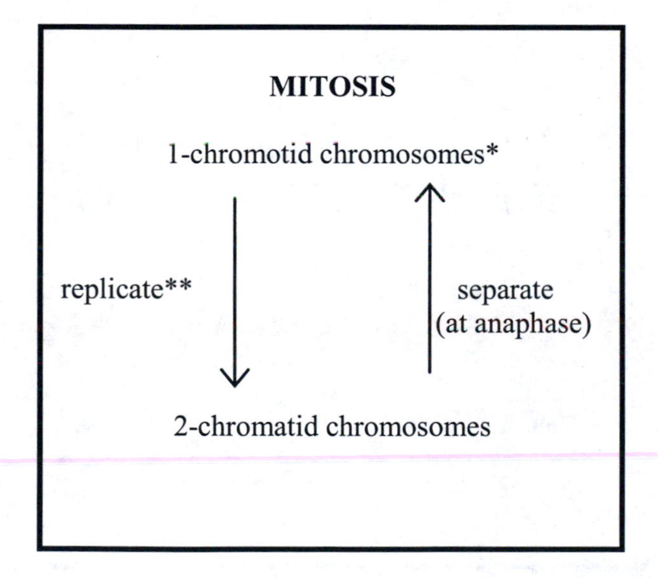

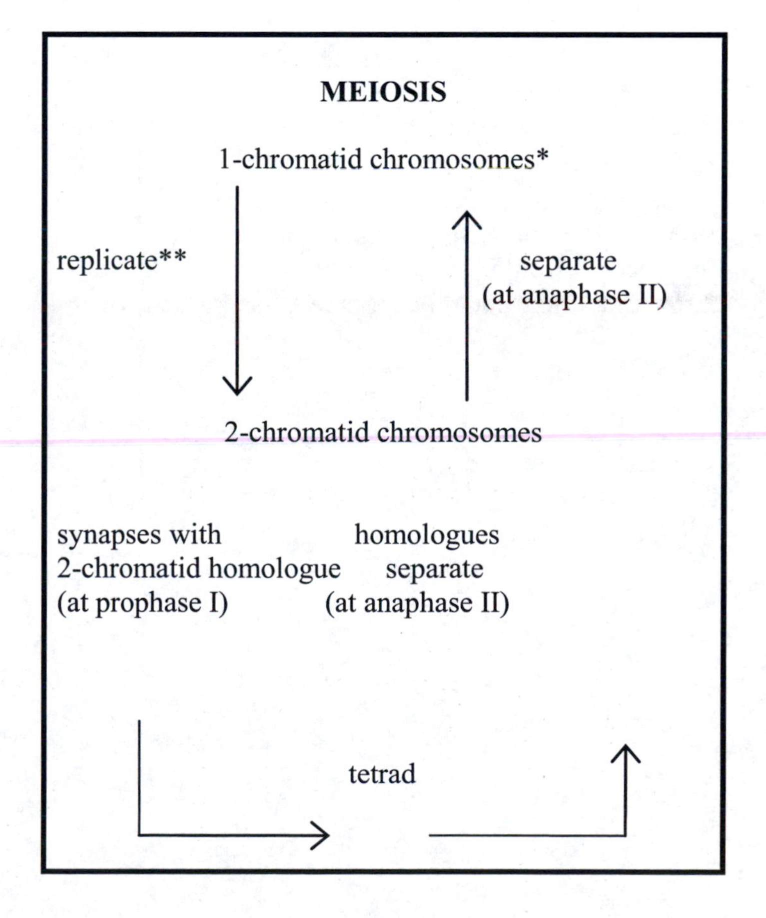

** Occurs during S Phase (preceding mitosis or meiosis)

* Distinct chromosomes are not actually distinguishable until mitosis begin (after S and G2) but each would have only one chromatid if they could be observed.

Figure 4.4: Number of Chromosomes per Cell

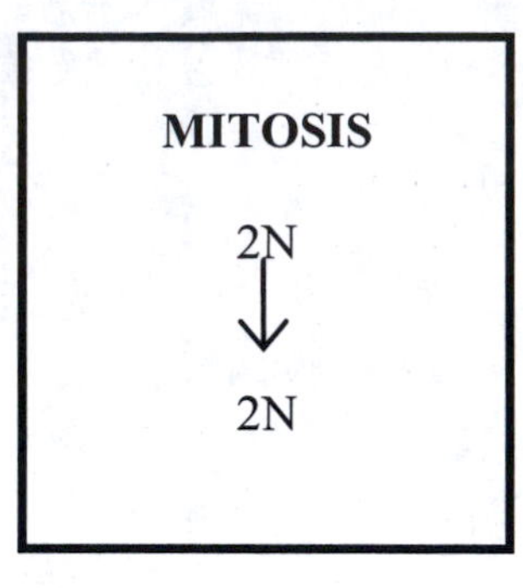

MEIOSIS

2N (1 strand of DNA (chromatin) per "chromosome"*)

↓ *replication***

2N (2 identical chromatids per chromosome)

Meiosis I
(Reduction Division

↓

1N (2 identical chromatids per chromosome)

Meiosis II

↓

1N (1 chromatid per chromosome)

+ 1N via fertilization

↓

2N (1 chromatid per chromosome)

* Distinct "chromosomes" are not actually distinguishable at this phase, but each would have only one chromatid if they could be observed.

** Occurs during G1, S and G2 of the cell cycle

Article 4.2: Spermatogenesis

In males, the formation of haploid sperm from the original diploid germ cell is called spermatogenesis. Primary spermatocytes begin the process by undergoing Meiosis I. The resulting two haploid cells are called secondary spermatocyes. These cells then undergo Meiosis II and produce four haploid spermatids that eventually differentiate into mature sperm cells.

The most striking change in each spermatocyte as it differentiates into a sperm is the formation of a flagellum, also called the tail. The movements of the tail require cellular energy and make the sperm motile. The sperm also has a head, which contains the nucleus that carries the chromosomes. The head and tail are separated by a midpiece containing one or more mitochondria, depending on the species, that produce the ATP required for mail movements. At the tip of the head is a special structure called the acrosome that contains proteolytic enzymes that help break down the protective outer layers surrounding the ovum.

Article 4.3: Oogenesis

Gamete production in males begins at puberty and continues throughout life but a female's total supply of eggs is determined by the times she is born, and the time span during which she releases them extends from puberty to menopause and occurs in the ovaries.

The first stage occurs during fetal development where the oogonia (diploid germ cells) rapidly multiply by mitosis and then enter a growth phase and lay in nutrient reserves. These cells are transformed into primary oocytes and become surrounded by a single layer of follicle cells. These cells enter Meiosis I and become stalled in late Prophase I. At puberty, a small number of these stalled cells are activated each month, however only one will continue complete Meiosis I. At the end of Meiosis I, two haploid cells of dissimilar size are produced: the smaller cell (almost no cytoplasm) is called the first polar body and the larger cell (virtually all the cytoplasm) is the secondary oocyte.

The first polar body usually undergoes Meiosis II producing two smaller polar bodies. However, the secondary oocyte arrests in Metaphase II and it is this cell that will be fertilized. If this cell is not fertilized it will deteriorate and be expelled during menstruation. However, it fertilization does occur, Meiosis II is quickly completed and a second polar body will also be created.

The unequal cytoplasmic division that occurs during oogenesis assures that a fertilized egg has ample nutrients for its seven-day journey from the fallopian tubes to the uterus. Once it implants in the uterus, the mother will supply the nutrients.

Supplemental Articles

For a student of Introductory Biology reading primary sources (journal articles that describe the scientists' research and results) is often a very difficult task. The nature of the research is often very specialized and thus very technical and requires advanced courses to understand the work. This is also the case in older research—what your textbook is based on. However, there are exceptions to the rule that beginning students cannot read primary sources and the following articles will expose you to two very important scientific papers. James Watson and Francis Crick wrote two papers on their discovery on the structure of the DNA molecule. In the first paper take note of its brevity; the actual article is only one page in the journal. Since they were in a rush to publish their findings and thus get credit for this important discovery, the first paper discusses what the molecule looks like (the details and supporting evidence would be published later) and only alludes to how DNA would replicate. The second paper completes this missing piece.

Take note of the end notes for both papers. Notice whose work is not credited but yet that data was crucial for the model to be built.

MOLECULAR STRUCTURE OF NUCLEIC ACIDS
A Structure for Deoxyribose Nucleic Acid

Nature, April 25, 1953 (pg. 737)

We wish to suggest a structure for the salt of deoxyribose nucleic acid (DNA). This structure has novel feature which are of considerable biological interest.

A structure for nucleic acid has already been proposed by Pauling and Corey[1]. They kindly made their manuscript available to us in advance of publication. Their model consists of three inter twisted chains, with the phosphates near the fibre axis, and the bases on the outside. In our opinion, this structure is unsatisfactory for two reasons: (1) We believe that the material which gives the X-ray diagrams is the salt, not the free acid. Without the acidic hydrogen atoms it is not clear what forces would hold the structure together, especially as the negatively charged phosphates near the axis will repel each other. (2) Some of the van der Waals distances appear to be too small.

Another three-chain structure has also been suggested by Fraser (in the press). In his model the phosphates are on the outside and the bases on the inside, linked together by hydrogen bonds. This structure as described is rather ill-defined, and for this reason we shall not comment on it.

We wish to put forward a radically different structure for the salt of deoxyribose nucleic acid. This structure has two helical chains each coiled round the same axis (see diagram—not shown). We have made the usual chemical assumptions, namely, that each chain consists of phosphate diester groups joining β-D-deoxyribofuranose residues with 3', 5' linkages. The two chains (but not their bases) are related by a dyad perpendicular to the fibre axis. Both chains follow right handed helices, but owing to the dyad the sequences of the atoms in the two chains run in opposite directions. Each chain loosely resembles Furberg's[2] model No. 1; that is, the bases are on the inside of the helix and the phosphates on the outside. The configuration of the sugar and the atoms near it is close to Furberg's 'standard configuration', the sugar being roughly perpendicular to the attached base. There is a residue on each chain every 3-4 A, in the *s*-direction. We have assumed an angle of 36° between adjacent residues in the same chain, so that the structure repeats after 10 residues on each chain, that is, after 34 A. The distance of a phosphorus atom from the fibre axis is 10 A. As the phosphates are on the outside, cations have easy access to them.

The structure is an open one, and its water content is rather high. At lower water contents we would expect the bases to tilt so that the structure could become more compact.

The novel feature of the structure is the manner in which the two chains are held together by purine and pyrimidine bases. The planes of the bases are perpendicular to the fibre axis. They are joined together in pairs, a single bases from one chain being hydrogen-bonded to a single bases from the other chain, so that the two lie side by side with identical *z*-co-ordinates. One of the pair must be a purine and the other a pyrimidine for bonding to occur. The hydrogen bonds are made as follows: purine position 1 to pyrimidine position 1; purine position 6 to pyrimidine position 6.

If is as assumed that the bases only occur in the structure in the most plausible tautomeric forms (that is, with the keto rather than the enol configurations) it is found that only specific pairs of bases can bond together. These pairs are: adenine (purine) with thymine (pyrimidine), and guanine (purine) with cytosine (pyrimidine).

In other words, if an adenine forms one member of a pair, on either chain, then on these assumptions the other member must be thymine; similarly for guanine and cytosine. The sequence of bases on a single chain does not appear to be restricted in any way. However, if only specific pairs of bases can be formed, it follows that if the sequence of bases on one chain is given, then the sequence on the other chain is automatically determined.

It has been found experimentally [3,4] that the ratio of the amounts of adenine to thymine, and the ratio of guanine to cytosine, are always very close to unity for deoxyribose nucleic acid.

It is probably impossible to build this structure with ribose sugar in place of the deoxyribose, as the extra oxygen atom would make too close a van der Waals contact.

The previously published X-ray data [5,6] on deoxyribose nucleic acid are insufficient for a rigorous test of our structure. So far as we can tell, it is roughly compatible with the experimental data, but it must be regarded as unproved until it has been checked against more exact results. Some of these are given in the following communications. We were not aware of the details of the results presented there when we devised our structure, which rests mainly thought not entirely on published experimental data and stereochemical arrangements.

It has not escaped out notice that the specific pairing we have postulated immediately suggests a possible copying mechanism for the genetic material.

Full details of the structure, including the conditions assumed in building it, together with a set of co-ordinates for the atoms, will be published elsewhere.

We are much indebted to Dr. Jerry Donohue, for constant advice and criticism, especially on interatomic distances. We have also been stimulated by a knowledge of the general nature of the unpublished experimental results and ideas of Dr. M. H. F. Wilkins, Dr. R. E. Franklin and their co-workers at King's College, London. One of us (J.D.W.) has been aided by a fellowship from the National Foundation for Infantile Paralysis.

Medical Research Council Unit for the
Study of the Molecular Structure of
Biological Systems,
Cavendish Laboratory, Cambridge
April 2

1. Pauling, L. and Corey, R. B., *Nature*, 171, 346 (1953); *Proc. U.S. Nat. Acad. Sci*, 39, 84 (1953).

2. Furberg, S., *Acts Chem. Scand.*, 6, 634 (1952).

3. Chargaff, E. for references see Zamenhof, S., Brawerman, G. and Chargaff, E., *Biochim et Biophys. Acta*, 9, 402 (1952).

4. Wyatt, G. R., *J. Gen. Physiol.*, 36, 201 (1952).

5. Astbury, W., Symp. Soc. Exp. Biol. 1, Nucleic Acid, 66 (Camb. Univ. Press, 1947).

6. Wilkins, M. H .F., and Randall, J. T., *Biochim et Biophys. Acta*, 10, 192 (1953).

GENETICAL IMPLICATIONS OF THE STRUCURE OF DEOXYRIBOSE NUCLEIC ACID

Nature, May 30, 1953 (pg. 964-966)

The importance of deoxyribonucleic acid (DNA) within living cells is undisputed. It is found in all dividing cells, largely if not entirely in the nucleus, where it is an essential constituent of the chromosomes. Many lines of evidence indicate that it is the carrier of a part of (if not all) the genetic specificity of the chromosomes and thus of the gene itself. Until now, however, no evidence has been presented to show how it might carry out the essential operation required of a genetic material, that of exact self-duplication.

We have recently proposed a structure[1] for the salt of deoxyribonucleic acid which, if correct, immediately suggests a mechanism for its self-duplication. X-ray evidence obtained by the workers at King's College, London[2], and presented at the same time, gives qualitative support to out structure and is incompatible with all previously proposed structures[3]. Though the structures will not be completely proved until a more extensive comparison has been made with the X-ray data, we now feel sufficient confidence in its general correctness to discuss its genetical implications. In doing so we are assuming that fibres of the salt of deoxyribonucleic acid are not artifacts arising in the method of preparation, since it has been shown by Wilkins and his co-workers that similar X-ray patterns are obtained from both the isolated fibres and certain intact biological materials such as sperm head and bacteriophage particles[2,4].

The chemical formula of deoxyribonucleic acid is now well established. The molecule is a very long chain, the backbone of which consists of a regular alternation of sugar and phosphate groups, as shown in Fig. 1. (not shown) To each sugar is attached a nitrogenous base, which can be of four different types. (We have considered 5-methyl cytosine to be equivalent to cytosine, since either can fit equally well into our structure.) Two of the possible bases—adenine and guanine—are purines, and the other two—thymine and cytosine—are pyrimidines. So far as is known, the sequences of bases along the chain is irregular. The monomer unit, consisting of phosphate, sugar and base, is known as a nucleotide.

The first feature of our structure which is of biological interest is that it consists not of one chain, but of two. These two chains are both coiled around a common fibre axis, as shown diagrammatically in Fig. 2. (not shown) It has often been assumed that since there was only one chain in the chemical formula there would only be one in the structural unit. However, the density, taken with the X-ray evidence[2], suggest very strongly that there are two.

The other biologically important feature is the manner in which the two chains are held together. This is done by hydrogen bonds between the bases, as shown schematically in Fig. 3. (not shown) The bases are joined together in pairs, a single bases from one chain being hydrogen-bonded to a single bases from the other. The important point is that only certain pairs of bases will fit into the structure. One member of a pair must be a purine and the other a pyrimidine in order to bridge between the two chains. If a pair consisted of two purines, for example, there would not be room for it.

We believe that the bases will be present almost entirely in their most probable tautomeric forms. If this is true, the conditions for forming hydrogen bonds are more restrictive, and the only pairs of bases possible are:

adenine with thymine;
guanine with cytosine,

The way in which these are joined together is shown in Figs. 4 and 5. (not shown) A given pair can be either way round. Adenine, for example, can occur on either chain; but when it does, its partner of the other chain must always be thymine.

This pairing is strongly supported by the recent analytical results[5], which show that for all sources of deoxyribonucleic acid examined the amount if thymine, and the amount of guanine close to the amount of cytosine, although the cross-ratio (the ratio of adenine to guanine) can vary from one source to another. Indeed, if the sequence of bases on one chain is irregular, it is difficult to explain these analytical results except by the sort of pairing we have suggested.

The phosphate-sugar backbone of our model is completely regular, but any sequence of the pairs of bases can fit into the structure. It follows that in a long molecule many different permutations are possible, and it therefore seems likely that the precise sequence of the bases is the code which carries the genetical information. If the actual order of the bases on one of the pair of chains were given, one could write down the exact order of the bases on the other one, because of the specific pairing. Thus one chain is, as it were, the complement of the other, and it is this feature which suggest how the deoxyribonucleic acid molecule might duplicate itself.

Previous discussions of self-duplication have usually involved the concept of a template, or mould. Either the template was supposed to copy itself directly or it was to produce a 'negative', which in its turn was to act as a template and produce the original 'positive' once again. In no case has it been explained in detail how it would do this in terms of atoms and molecules.

Now our model for deoxyribonucleic acid is, in effect, a *pair* of templates, each of which is complementary to the other. We imagine that prior to duplication the hydrogen bonds are broken, and the two chains unwind and separate. Each chain then acts as a template for the formation on to itself of a new companion chain, so that eventually we shall have *two* pairs of chains, where we only had one before. Moreover, the sequence of the pairs of bases will have been duplicated exactly.

A study of our model suggests that this duplication could be done most simply if the single chain (or the relevant portion of it) takes up the helical configuration. We imagine that at this stage in the life of the cell, free nucleotides, strictly polynucleotide precursors, are available in quantity. From time to time the base of a free nucleotide will join by hydrogen bonds to one of the bases on the chain already formed. We now postulate that the polymerization of these monomers to form a new chain is only possible if the resulting chain can form the proposed structure. This is plausible, because steric reasons would not allow nucleotide 'crystallized' on to the first chain to approach one another in such a way that they could be joined together into a new chain, unless they were those nucleotides which were necessary to form our structure. Whether a special enzyme is required to carry out the polymerization or whether the single helical chain already formed acts effectively as an enzyme, remains to be seen.

Since the two chains in our model are intertwined, it is essential for them to untwist if they are to separate. As they make one complete turn around each other in 34 A., there will be about 150 turns per million molecular weight, so that whatever the precise structure of the chromosome a considerable amount of uncoiling would be necessary. It is well known from microscope observation that much coiling and uncoiling occurs during mitosis, and though this is on a much larger scale it probably reflects similar processes on a molecular level. Although it is difficult at the moment to see how these processes occur without everything getting tangled, we do not feel that this objection will in insuperable.

Our structure, as described[1], is an open one. There is room between the pair of polynucleotide chains (see Fig.2) (not shown) for a polypeptide chain to wind around the same helical axis. It may be significant that the distance between adjacent phosphorous atoms, 7-1 A., is close to the repeat of a fully extended polypeptide chain. We think it probable that in the sperm head, and in artificial nucleoproteins, the polypeptide chain occupies this position. The relative weakness of the second layer-line in the published X-ray pictures[3a, 4] is crudely compatible with such an idea. The function of the protein might well be to control the coiling and uncoiling, to assist in holding a single polynucleotide chain in a helical configuration, or some other non-specific function.

Our model suggests possible explanations for a number of other phenomena. For example, spontaneous mutation may be due to a base occasionally occurring in one of its less likely tautomeric forms. Again, the pairing between homologous chromosomes at meiosis may depend on pairing between specific bases. We shall discuss these ideas in detail elsewhere.

For the moment, the general scheme we have proposed for the reproduction of deoxyribonucleic acid must be regarded as speculative. Even if it is correct, it is clear from what we have said that much remains to be discovered in detail. What are the polynucleotide precursors? What makes the pair of chains unwind and separate? What is the precise role of the protein? Is the chromosome one long pair of deoxyribonucleic acid chains, or does it consist of patches of the acid joined together by protein?

Despite these uncertainties we fell that our proposed structure for deoxyribonucleic acid may help to solve one of the fundamental biological problems—the molecular basis of the template needed for genetic replication. The hypothesis we are suggesting is that the template is the pattern of bases formed by one chain of the deoxyribonucleic acid and that the gene contains a complementary pair of such templates.
One of us (J. D. W.) has been aided by a fellowship from the National Foundation for Infantile Paralysis (U. S. A.)

1. Watson, J. D. and Crick, F. H. C., *Nature*, 171, 737 (1953).

2. Wilkins, M. H. F., Stokes, A. R. and Wilson, H. R., *Nature*, 171, 738 (1953). Franklin, R. E., and Gosling, R. G., *Nature*, 171, 740 (1953).

3(a). Astbury, W. T., Symp. No. 1 Soc. Exp. Biol., 66 (1947). (b) Furberg, S., *Acta Chem. Sound.*, 6, 634 (1952). (c) Pauling, L., and Corey, R. B., *Nature*, 171, 346 (1953). (d) Fraser, R. D. B., (in preparation).

4. Wilkins, M. H. F. and Randall, J. T., *Biochim et Biophys. Acta*, 10, 192 (1953)

5. Chargaff, E. for ref. see Samenhof, S., Brawerman, G. and Chargaff, E., *Biochim et Biophys. Acta*, 9, 402 (1952). Wyatt, G. R., *J. Gen. Physiol.,* 36, 201 (1952).

After Class Work

1. A DNA molecules has 200 base pairs (400 bases). If 90 of the bases are thymine, how many bases are there of the other three types?

   a. adenine = ________

   b. guanine = ________

   c. cytosine = ________

2. The pictured segment of DNA is ready to replicate. The strands have uncoiled, the hydrogen bonds have broken and the two strands have come apart as if they were unzipped. The two separate strands may look like the following diagram. Draw in the complementary strands to show what this diagram would look like following replication.

Strand M:	Strand L:
5’	3’
G	C
G	C
T	A
G	C
T	A
A	T
A	T
T	A
T	A
3’	5’

3. Complete the following table.

DNA Coding Stand	5’CCACATTAA3’	5’GATCGATAT3’	5’AGACACGGA3’
DNA Template			
mRNA			
tRNA			
Amino Acids			

4. Write the codons for a polypeptide constructed as follows:

   start, glutamic acid, tryptophan, histidine

5. What would happen to the chromosome number in gametes if they were formed by the mitotic process instead of the meiotic process?

6. How are the chromosomes in the zygote related to the parents of the male and females whose gametes united to form it?

7. What is the relationship between a chromosome and a chromatid?

8. Why does crossing over occur only during synapsis?

9. In a model, the chromosomes from the male parent are blue and those from the female parent are white. (Since there are 6 chromosomes, 3 are blue and 3 are white.) Does this mean that after meiosis, a gamete will have 3 white chromosomes and another gamete will have 3 blue?

10. What distinguishes the second stage of meiosis from mitosis?

11. What might happen to a gene during crossing over?

12. What would happen to the traits if homologous chromosomes did not come together as tetrads?

13. Complete this chart to distinguish meiosis from mitosis.

Description	Mitosis	Meiosis
Number of Divisions Required		
Type of Chromosomes that Separate during Anaphase		
No. of Chromosomes in Daughter Cells		
Type of Cell Process Creates		

14. Match the item on the lettered list that best matches each item on the numbered list. Do not use any letter more than once.

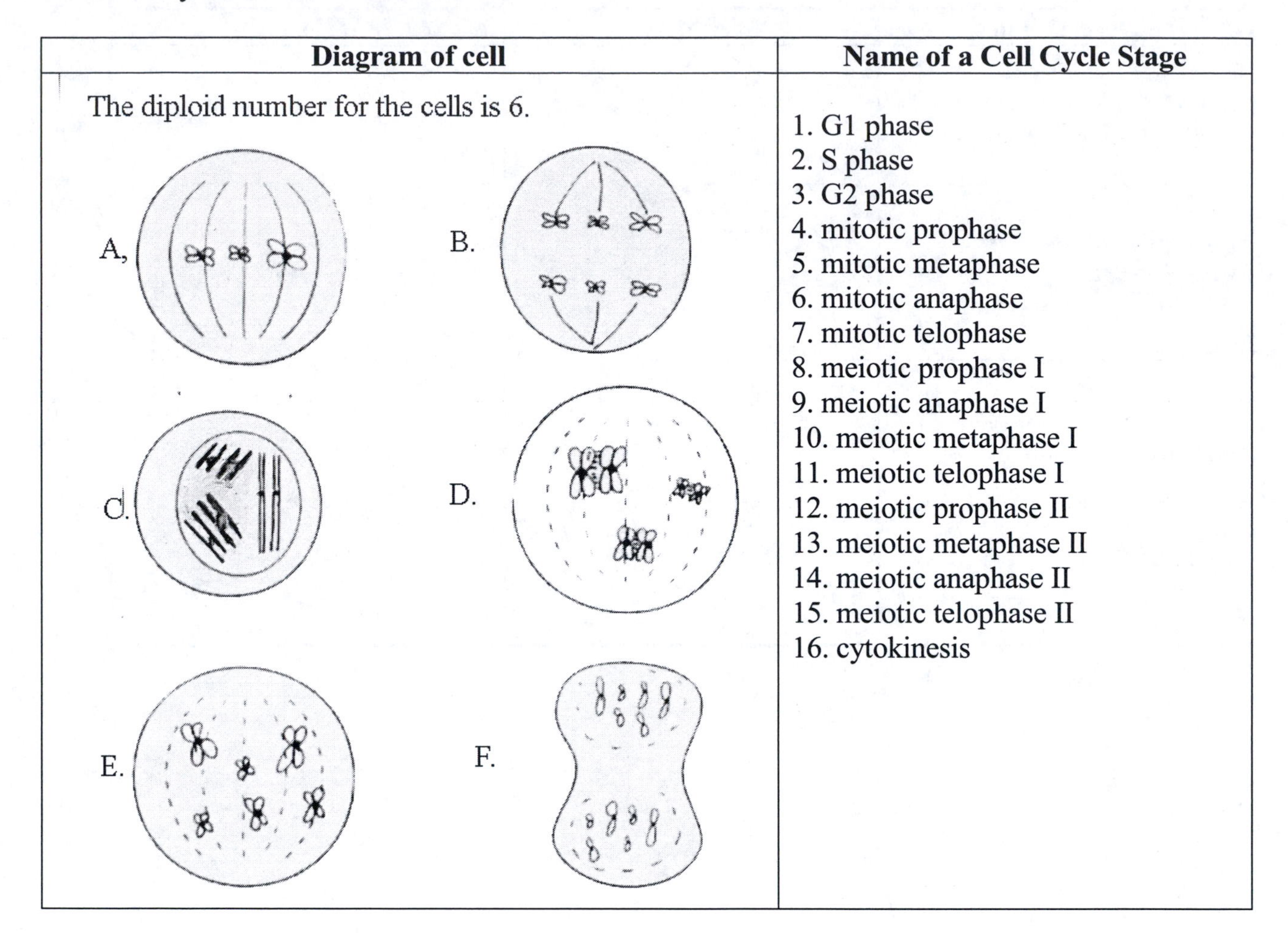

Diagram of cell	Name of a Cell Cycle Stage
The diploid number for the cells is 6. A, B. C. D. E. F.	1. G1 phase 2. S phase 3. G2 phase 4. mitotic prophase 5. mitotic metaphase 6. mitotic anaphase 7. mitotic telophase 8. meiotic prophase I 9. meiotic anaphase I 10. meiotic metaphase I 11. meiotic telophase I 12. meiotic prophase II 13. meiotic metaphase II 14. meiotic anaphase II 15. meiotic telophase II 16. cytokinesis

Answers

1. a. 90　　　　b. 110　　　　c. 110

2. Complimentary strand for M: 3'CCACATTAA5'
   Complimentary strand for L: 5'GGTGTAATT3'

3.

DNA Coding Strand	**5'CCACATTAA3'**	**5'GATCGATAT5'**	**5'AGACACGG3'**
DNA Template	3'GGTGTAATT5'	3'CTAGCTATA5'	3'TCTGTGCCT5'
mRNA	5'CCACAUUAA3'	5'GAUCGAUAU3'	5'UGACACGGA3'
tRNA	GGUGUAAUU	CUAGCUAUA	UCUGUGCCU
Amino Acids	proline histidine	aspartate arginine tyrosine	arginine histidine glycine

4. Start = AUG
   Glutamic Acid = GAA or GAG
   Tryptophan = UGG
   Histidine = CAU or CAC

5. In each generation the chromosome number would double.

6. The new individual will have ½ the total number from each of the female's parents and ½ the total number from each of the male's parents, if chromosomes from the parents are distributed evenly and there is no crossing over.

7. A chromatid is one part of a chromosome that has replicated and is still attached to the other chromatid at the centromere.

8. Only during synapsis are the joined chromatids of homologous chromosomes close enough to twist around one another and exchange parts.

9. No. Since chromosomes are distributed randomly any combination of blue and white may occur.

10. There is no replication of DNA preceding the second stage of meiosis.

11. It could end up as part of a different chromosome.

12. There would be a random jumble of chromosomes separating in anaphase and more than one gene (chromosome) could be inherited for a given trait.

13.

Description	Mitosis	Meiosis
Number of Divisions Required	one	two
Type of Chromosomes that Separate during Anaphase	chromatids	homologous
No. of Chromosomes in Daughter Cells	diploid (2N)	haploid (N)
Type of Cell Process Creates	body (somatic) cells	gametes eggs—female; sperm—males

14. A. 13
B. 9
C. 8
D. 10
E. 4
F. 7

Sample Multiple Choice Questions

For mitosis and meiosis you must be able to recognize diagrams of the various stages and structures and processes.

1. Uracil forms a complementary pair with _____ in RNA and _____ in DNA.
   a. adenine; adenine
   b. adenine; thymine
   c. thymine; thymine
   d. uracil; adenine
   e. adenine; uracil

2. RNA synthesis is also known as
   a. elongation.
   b. initiation.
   c. termination.
   d. translation.
   e. transcription.

3. Introns in pre-mRNA are known to
   a. code for specific protein regions.
   b. undergo excision, whereby they are spliced out of the message.
   c. be able to move within the mRNA, thereby giving rise to new exon combinations.
   d. protect pre-mRNA from enzyme degradation.

4. If the DNA sequence is GCA, then the resulting amino acid will be
   a. CGU.
   b. CGT.
   c. alanine.
   d. arginine.

5. The main reason scientists thought that proteins, rather than DNA, were the carriers of genetic material in the cell was
   a. their presence within the nucleus.
   b. their abundance within the cell.
   c. the large number of possible amino acid sequences.
   d. their ability to self-replicate within the cytoplasm.
   e. their ability to be exported with the cell.

6. Which of the following nucleotide sequences represents the complement to the DNA strand:

   5’ AGATCCG 3’

   a. 5’ AGATCCG 3’
   b. 3’ AGATCCG 5’
   c. 5’ CTCGAAT 3’
   d. 3’ CTCGAAT 5’
   e. 3’ TCTAGGC 5’

7. Which of the following adds new nucleotides to a growing DNA chain?
   a. DNA polymerase
   b. DNA helicase
   c. primase
   d. RNA polymerase
   e. Ligase

8. A cell is in metaphase if the
   a. chromosomes are visible as thread-like structures.
   b. nuclear envelope is beginning to disintegrate.
   c. chromosomes are aligned on the equatorial plane of the cell.
   d. chromosomes are separated into distinct groups and are moving to opposite poles.

9. _____ contain identical DNA sequences and are held together by _____.
   a. Daughter chromosomes, hydrogen bonding
   b. Daughter chromosomes, ionic bonding
   c. Daughter chromatids, spindle fibers
   d. Sister chromosomes, histone proteins
   e. Sister chromatids, centromers

10. The zygote contains the _____ complement of chromosomes.
    a. haploid
    b. diploid
    c. polyploidy
    d. spermatogenesis
    e. oogenesis

11. Translocation occurs when
    a. a part of a chromosomes breaks off and attaches to a non-homologous chromosome.
    b. part of a chromosomes breaks off and attaches to a homologous chromosome.
    c. crossing-over events occur.
    d. genes move from one area on a chromosome to another area of the same chromosome.
    e. a Y chromosome replaces an X chromosome in a female.

12. Persons have XXY karyotype are nearly normal males but produce few or no sperm. They have _____ Syndrome.
    a. Turner’s
    b. Klinefelter
    c. Downs
    d. metafemale

Answers to Multiple Choice Questions

A. Uracil always binds to adenine. Guanine and cytosine are always paired together. The exception is what adenine will bind to. In DNA, adenine pairs with thymine; but when making with RNA, adenine will pair to uracil.

E. DNA acts as the template to make all three types of RNA molecules. Translation (D) is converting the RNA into amino acids (building blocks of proteins). Elongation (A) is part of translation where the polypeptide chains grows through the addition of amino acids. Initiation (B) is the start of translation where the components all come together and (C) termination is when the stop codon is reached and the components "fall apart".

B. Introns are intervening sequences in eukaryotic DNA and are sometimes referred to as "junk DNA". They are transcribed from the DNA into the pre-mRNA and are removed before the molecule leaves the nucleus. The remaining segments are called exons (A) because they contain the sequences that will be expressed (translated). Choice C refers to Barbara McClintock's work on jumping genes (transposons) where some DNA segments slip into and out of different locations in the DNA molecule. Choice D refers to the further modification to the pre-mRNA where caps and tails are added to the molecule.

D. In order to use the Genetic Dictionary you must be in the language of the messenger. Therefore you must convert the DNA code of GCA to mRNA which will be CGU. This triplet is now called a codon and when looked up on the chart will result in the amino acid arginine. If you came up with C as the answer it is because you used the DNA code with the dictionary. Choices A and B are nucleotide sequences.

C. The Tetranucleotide Theory said that because proteins were composed of 20 different amino acids and DNA was made up of only 4 different nucleotides, it was easier to make more unique combinations with 20 building blocks rather than 4.

E. DNA is a double-stranded molecule where each stranded is composed of nucleotides. The bases will be on the inside on the molecule and will follow the complementary base pairing rules where adenine hydrogen bonds to thymine, guanine to cytosine. The backbone (alternating sugar and phosphate) are not only parallel to each other but they are upside down to each other. This is called anti-parallel, so each strand has a different end of the sugar exposed (3' on one, 5' on the other).

A. DNA polymerase reads the template, adds the nucleotides and also proofreads the new strand. Helicase unwinds and unzippers the original molecule. Primase brings in a small number of RNA nucleotides so that the first DNA will have something to "grab onto". RNA polymerase is involved in transcription. Ligase seals the DNA fragments that were created from the original.

C. This is the correct description of metaphase. Choice A is interphase, B is prophase (also the chromatin condenses to form chromosomes), and D is anaphase. The last state is telophase were the new cells are created.

E. Choices A and B make no sense. Spindle fibers are the fibers generated at the poles of the cells and are used to move the chromatids to the poles. Histones are proteins that help stabilize the DNA in eukaryotic cells—the DNA wraps around these proteins and makes nucleosomes.

B. The zygote is the fertilized egg that is the result of fertilization. A haploid (N) sperm and haploid ova join and a 2N (diploid) cell is created. Spermatogenesis is the process of creating sperm and oogenesis makes eggs. These processes are using meiosis. A polyploid has complete extra sets of chromosomes (3N, 4N, etc.)

A. Choices B and C are describing the same thing, crossing over, which is a good thing because it leads to increased genetic diversity of the off spring. Translocation is bad because they resulting gametes will either have extra chromosomal material or will be missing sections.

B. Turners are XO females. Downs is most frequently caused by trisomy of chromosome 21. The metafemale is XXX.

UNIT 5: GENETICS

Outline

TOPIC	SOLOMON
I. From Previous Units	
A. Chromosomes	
1. Structure	
2. Autosomal and Sex	
B. Meiosis	
1. Process	
2. Crossing Over	
II. Mendelian Patterns of Inheritance	
A. Blending Theory	227
B. Mendel's Work	228
C. Monohybrid Crosses	229-232
1. Modern Terminology	
a. Genotype and Phenotype	
b. Locus and Alleles	
c. Dominant and Recessive	
d. Homozygous and Heterozygous	
2. Generations	
3. Punnett Squares	236-237
a. Probability	
4. Law of Segregation	
5. Test Cross	
D. Dihybrid Crosses	233-234
1. Law of Independent Assortment	
2. Gamete Production	
3. Linkage	238, 241
E. Human Disorders Inherited by Mendelian Rules	338, 347-350
1. Pedigree Charts	
2. Autosomal Recessive	
a. Cystic Fibrosis	
b. Tay Sachs	
c. Phenylketonuria	
d. Albinism	
3. Autosomal Dominant	
a. Huntingtons	

TOPIC	SOLOMON
IV. Other Inheritance Patterns	
A. Wild types vs. mutants	
B. Incomplete Dominance	244
1. Human Disorders	
a. Sickle Cell Anemia and Trait	
C. Multiple Alleles and Codominance	245-246
1. Blood Donation	
2. Rh Factor	
D. Polygenic Inheritance	247
E. Pleiotrophy	246
1. Human Disorders	
a. Marfan's Syndrome	
b. Sickle Cell Anemia	
c. Cystic Fibrosis	
F. Epistasis	246
G. Environmental Effects	248
1. Himalayan Rabbits	
2. Thalidomide Babies	
H. Chromosomal Theory of Inheritance	255
1. Thomas Hunt Morgan	
I. Sex Chromosomes	240
1. X or Sex Linked	
a. Human Disorders	
(i) Hemophilia	250
(ii) Red/Green Colorblindness	
J. Types of Chromatin	242-244
1. Euchromatin Material	
2. Heterochromatin Material	
a. Barr Bodies	
K. Sex Influenced Traits	
L. Parental Imprinting	
a. Fragile-X Syndrome	
V. Visual Supplements	
A. Thinkwell CD-ROM	
1. Molecular Genetics	
a. Eukaryotic Genomic Organization	
VII. Exam	

Objectives

1. Apply the chromosomal events of meiosis to Mendel's Laws of Segregation and Independent Assortment.

2. Define the following terms: alleles, genes, loci, heterozygous, homozygous, recessive, dominant, genotype, phenotype and wild type.

3. Apply the Laws of Probability or use Punnett Squares to the distribution of hereditary traits in monohybrid crosses and assign proper genotypes and phenotypes to the parents predicted offspring.

4. Determine the gametes that can be produced from a dihybrid cross. Solve dihybrid problems.

5. Solve genetic problems involving incomplete dominance, ABO blood types and sex-linkage.

6. Explain the relationship between recessive phenotypes, incomplete dominance and non-functioning and/or missing proteins.

7. Given an individual's blood type be able to determine what type of blood that person can receive and to whom they can donate.

8. Define pleiotrophy, epistasis and polygenic inheritance. Give examples for each inheritance pattern.

9. Explain the role the environment has on gene expression.

10. In regard to the Chromosome Theory of Inheritance, be able to define the theory and the evidence that led to its development.

11. Explain sex-influenced traits.

12. What is the mode of inheritance, major characteristics and treatments (if applicable) for the following genetic disorders: Cystic Fibrosis, Tay-Sachs, Phenylketonuria, albinism, Rh factor, Huntington's, Sickle Cell Anemia, Sickle Cell Trait, Marfan's Syndrome, Hemophilia, Red/ Green Colorblindness and Fragile X Syndrome.

13. Explain how a mutation might change the phenotype and why mutation do not always result in a recessive phenotype.

Article 5.1: Mendelian Misconceptions

BioQuest Notes, May 1999

- Dominant traits are "stronger" and "overpower" the recessive trait
- Dominant traits are more likely to be inherited
- Dominant traits are more "fit" or more adaptive in terms of natural selection. Any recessive adaptive mutant trait will eventually evolve to become dominant
- Dominant traits are more prevalent in the population
- Dominant traits are "better"
- "Wild-type" or "natural" traits are dominant, whereas mutants are recessive
- Male or masculine traits are dominant

Article 5.2: Human Blood Groups

The ABO blood groups are based on the presence or absence of two glycoproteins, called type A and type B. Depending on which of these a person inherits, their blood type will be one of the following: A, B, AB or O. The O blood groups, which has neither glycoprotein, is the most common ABO blood groups in white, black, and Asian Americans; with AB being the least frequent.

Unique to the ABO blood groups is the presence in the plasma of preformed antibodies. (Develop at birth upon exposure to bacteria that carry similar antigens—glycoproteins—on their surfaces.)

Blood Type	**Glycoproteins Present**	**Antibodies**
A	A	Anti-B
B	B	Anti-A
AB	A and B	None
O	none	Anti-A and Anti-B

Donor	**Recipient**			
	A	**B**	**AB**	**O**
A	yes	no	yes	no
B	no	yes	yes	no
AB	no	no	yes	no
O	yes	yes	yes	yes

The Rh blood typing system is so named because one of the Rh antigens (called D) was originally identified in the rhesus monkey. Approximately 85% of Americans are Rh+, meaning that their red blood cells carry the antigen. Unlike the ABO system, anti-Rh antibodies do not form unless the immune system of an Rh- person is directly exposed to the antigen (through transfusion, birthing a baby, abortion, miscarriage).

Article 5.3: What is Thalidomide?

Thalidomide Victims Association of Canada
http://www.thalidomide.ca/english/wit.html

Chemie Grünenthal synthesized thalidomide in West Germany in 1953. It was marketed (available to patients) from October 1, 1957 (West Germany) into the early 1960's. Thalidomide was present in at least 46 countries under many different brand names. Thalidomide became available in "sample tablet form" in Canada in late 1959. It was licensed for prescription use on April 1, 1961. Although Thalidomide was withdrawn from the West German and United Kingdom markets by December 2, 1961, it remained legally available in Canada until March 2, 1962. Incredulously Thalidomide was still available in some Canadian pharmacies until mid-May 1962. Thalidomide, was hailed as a "wonder drug" that provided a "safe, sound sleep". Thalidomide was a sedative that was found to be effective when given to pregnant women to combat many of the symptoms associated with morning sickness. It was not realized that Thalidomide molecules could cross the placental wall affecting the fetus until it was too late.

Thalidomide was a catastrophic drug with tragic side effects. Not only did a percentage of the population experience the effects of peripheral neuritis, a devastating and sometimes irreversible side effect, but Thalidomide became notorious as the killer and disabler of thousands of babies. When Thalidomide was taken during pregnancy (particularly during a specific window of time in the first trimester), it caused startling birth malformations and death to babies. Any part of the fetus was in development at the time of ingestion could be affected.

For those babies who survived, birth defects included: deafness, blindness, disfigurement, cleft palate, many other internal disabilities, and of course the disabilities most associated with Thalidomide: phocomelia (characterized by absence of a limb(s) or limbs attached to the trunk).

Current uses for thalidomide as a drug of last resort include:

- Dermatoses
- HIV-ulcers
- HIV-related conditions
- Cancer
- Leprosy

Article 5.4: Fragile X Syndrome

Fragile X Syndrome is one of the most common genetic causes of mental retardation, second only to Downs Syndrome. If affects about one in 1,500 males and one in 2,500 females and is seen in all ethnic groups. It is called Fragile X Syndrome because its diagnosis used to be dependent upon observing an X chromosome whose tip is attached to the test of the chromosome by a thin thread.

The inheritance pattern of Fragile X Syndrome is not like any other pattern we have studied. The chance of being affected increases in successive generations almost as if the pattern of inheritance switches from being a recessive one to a dominant one. Then too, an unaffected grandfather can have grandchildren with the disorder; in other words he is a carrier for an X-linked disease. This is contrary to normality since we have learned that only females can be carriers of X-linked traits.

In 1991, the DNA sequence at the fragile site was isolated and found to have trinucleotide repeats. The base triplet, CCG, was repeated over and over again. There are about 6-50 copies of this repeat in normal persons and over 230 copies in persons with Fragile X Syndrome. Carrier males have what is now termed a permutation; they have between 50 and 230 copies of the repeat and no symptoms. Both daughters and sons receive the permutation but only the daughters pass on the full mutation; that is 230 copies of the repeat. It is unknown what causes the difference between males and females.

This type of mutation, called by some a dynamic mutation because it changes, and by others an expanded trinucleotide repeat because the number of triplet copies increases, is now known to characterize other conditions. With Huntington's, the age of onset of the disorder is roughly correlated with the number of repeats, and the disorder is more likely to have been inherited from the paternal parent. In keeping with Mendel's findings, we would expect the sex of the parent to play no role in inheritance. The present exceptions have led to the genomic imprinting hypothesis, that the sperm and egg carry chromosomes that have been "imprinted" differently. Imprinting is believed to occur during gamete formation and thereafter the genes expressed one way if donated by the father and another if donated by the mother. Perhaps when we discover why more repeats are passed on by one parent rather than the other, we will discover the cause of so called genomic imprinting.

What might cause repeats to occur in the first place? Something must go wrong during DNA replication prior to cell division. The cell must find it difficult to correctly copy triplets that contain CG or GC combinations because all of the repeats so far noted have such combination. What difficulty arises because of triplet repeats? DNA codes for cellular proteins and the presence of repeats undoubtedly leads to nonfunctioning or malfunctioning proteins.

Figure 5.3: Fragile X Syndrome Chart

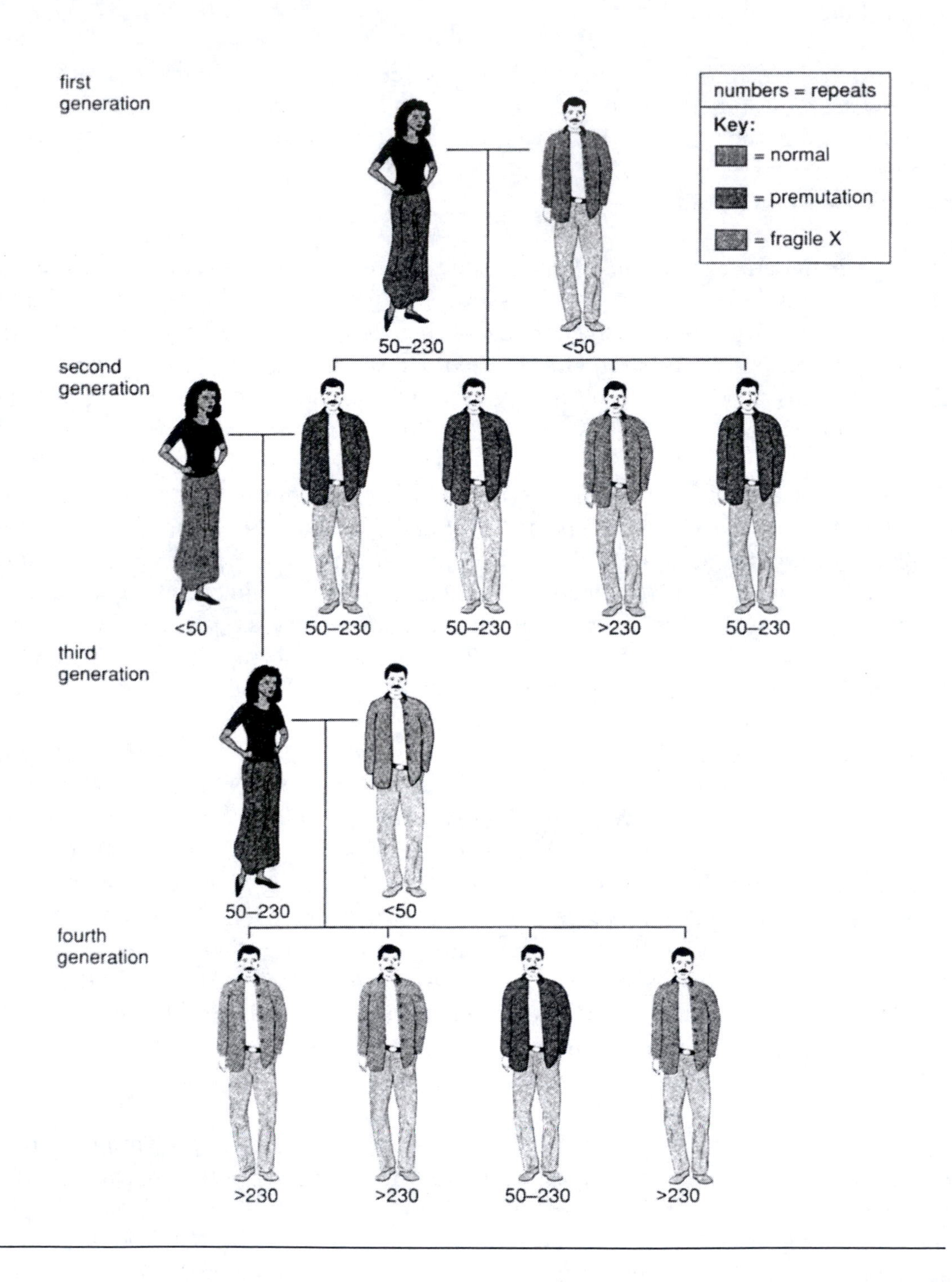

Worksheet 5.1: Genetic Disorders

DISORDER	MODE OF TRNASMISSION	PHENOTYPIC EXPRESSION	TREATMENT	OTHER INFORMATION
Cystic Fibrosis				
Tay Sachs				
Phenylketonuria (PKU)				
Albinism				

DISORDER	MODE OF TRNASMISSION	PHENOTYPIC EXPRESSION	TREATMENT	OTHER INFORMATION
Rh Factor				
Huntington's				
Sickle Cell Anemia & Trait				
Marfan's Syndrome				

DISORDER	**MODE OF TRNASMISSION**	**PHENOTYPIC EXPRESSION**	**TREATMENT**	**OTHER INFORMATION**
Hemophilia				
Red/Green Colorblindness				
Fragile X				

Worksheet 5.2: Pedigree Charts

Constructing a Human Pedigree Chart:

Data for 21 persons in a family pedigree chart was collected and recorded on the chart on the next page. The following traits were investigated, with the indicated patterns of inheritance.

Widow's Peak is a autosomal dominant trait (W), and is dominant over straight or continuous hairline (ww).

Hitch-hiker's thumb is when your thumb bends backwards when held vertical with the fingers in a fist. It is an autosomal recessive trait (tt). Straight thumb is dominant (T).

Hair: The protein in hair is an example of incomplete dominance where curls (I^1) incompletely dominant over straight (I^2). The intermediate (heterozygote) phenotype is wavy hair.

Blood typing is an example of multiple alleles and codominance. Three alleles exist in the population: I^A, I^B, and i. I^A and I^B are both dominant over the i allele; however, when I^A and $I^B$are both present, they are equally expressed (codominant) and result in a blood type of AB. Blood type O results from the recessive ii.

Color blindness is an X-linked, or sex-linked recessive trait, meaning it is carried on the X chromosome. In females, both recessives must be present to have the disorder; however, in males, because they carry only one X chromosome, only one recessive is required to show the disorder. The normal vision allele is denoted as X^B and the defective or colorblind allele as X^b.

PERSON	WIDOW PEAK	HITCH THUMB	HAIR	BLOOD TYPE	COLORBLIND
Mary Sycamore	Y	Y	S	A	
John Sycamore	Y			A	
Eleanor Redwood		Y		O	Y
Harry Redwood	N	Y	S		
Ed Sycamore	Y	Y	S	O	N
Sue Sycamore	N	N	S	O	Y
Joe Sycamore	N			A	
Ann Redwood					
Fred Redwood	N		S	A	
Alice Redwood	Y		W	B	
Phil Birch	N	Y	S	O	
Kim Birch		Y	C	O	Y
Jim Sycamore	N	Y	S	AB	N
Rick Sycamore	N		S	AB	
Sally Sycamore	N	Y	S	AB	
Laura Birch	N				N
Stan Birch	Y				
Sarah Birch	N				
Mike Sycamore		Y		A	N
Diane Sycamore		Y		AB	Y
Steve Sycamore		N		B	N

For all the traits except hair, a "Y" on the above chart indicates that the person is known to have that particular trait. An "N" indicates that the person is known not to have that particular trait. For hair, "S" is straight, "W" is wavy and "C" is curly. A blank indicates that it is not known what trait is expressed.

Your goal is to figure out all the missing traits. Treat each trait independently. Label each chart clearly, two traits can be completed on each chart. Place the genotypes within the circles and squares and write the phenotypes below for each person below their circle or square.

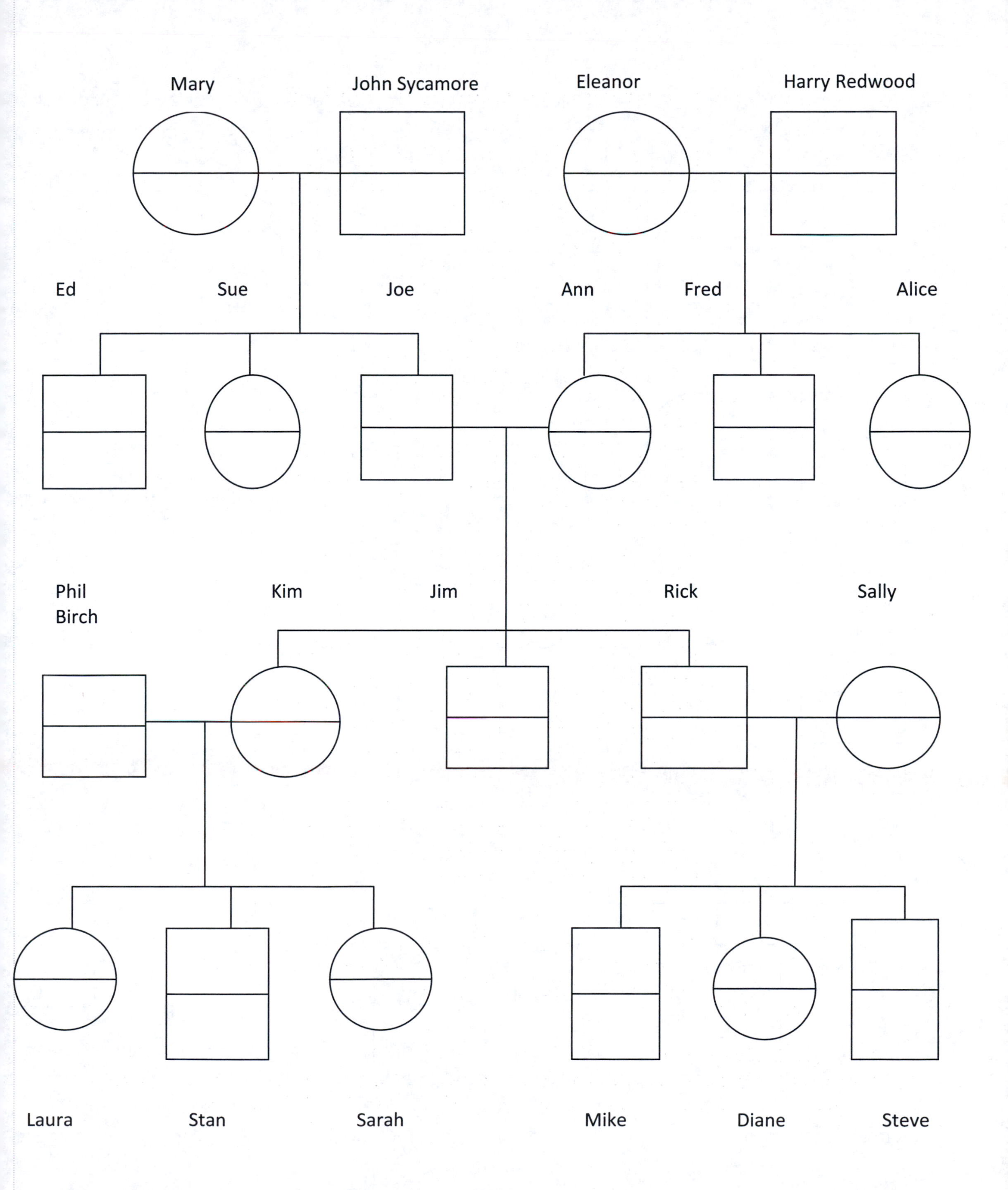
Mary
John Sycamore
Eleanor
Harry Redwood
Ed
Sue
Joe
Ann
Fred
Alice
Phil
Birch
Kim
Jim
Rick
Sally
Laura
Stan
Sarah
Mike
Diane
Steve

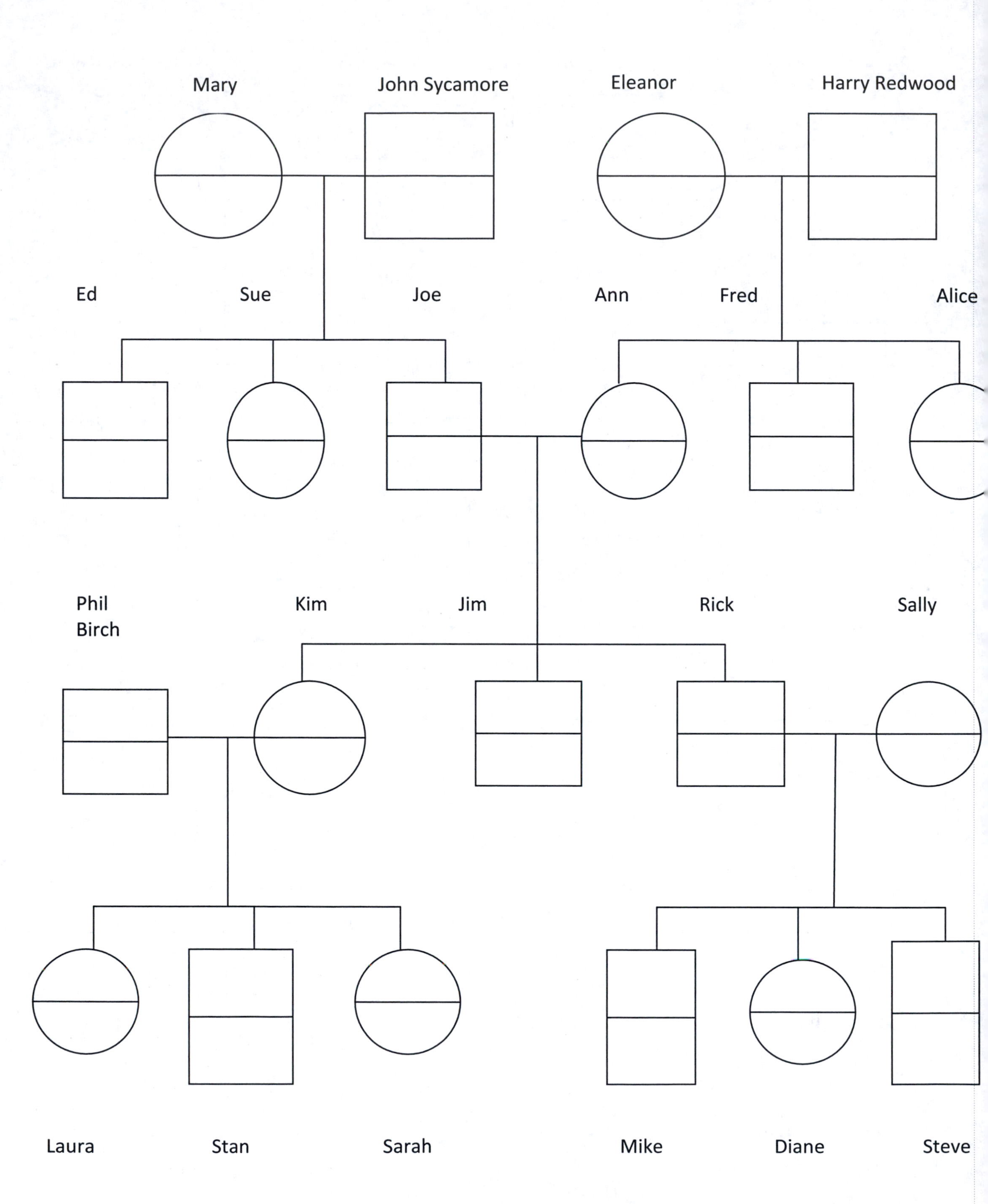
Mary
John Sycamore
Eleanor
Harry Redwood
Ed
Sue
Joe
Ann
Fred
Alice
Phil
Birch
Kim
Jim
Rick
Sally
Laura
Stan
Sarah
Mike
Diane
Steve

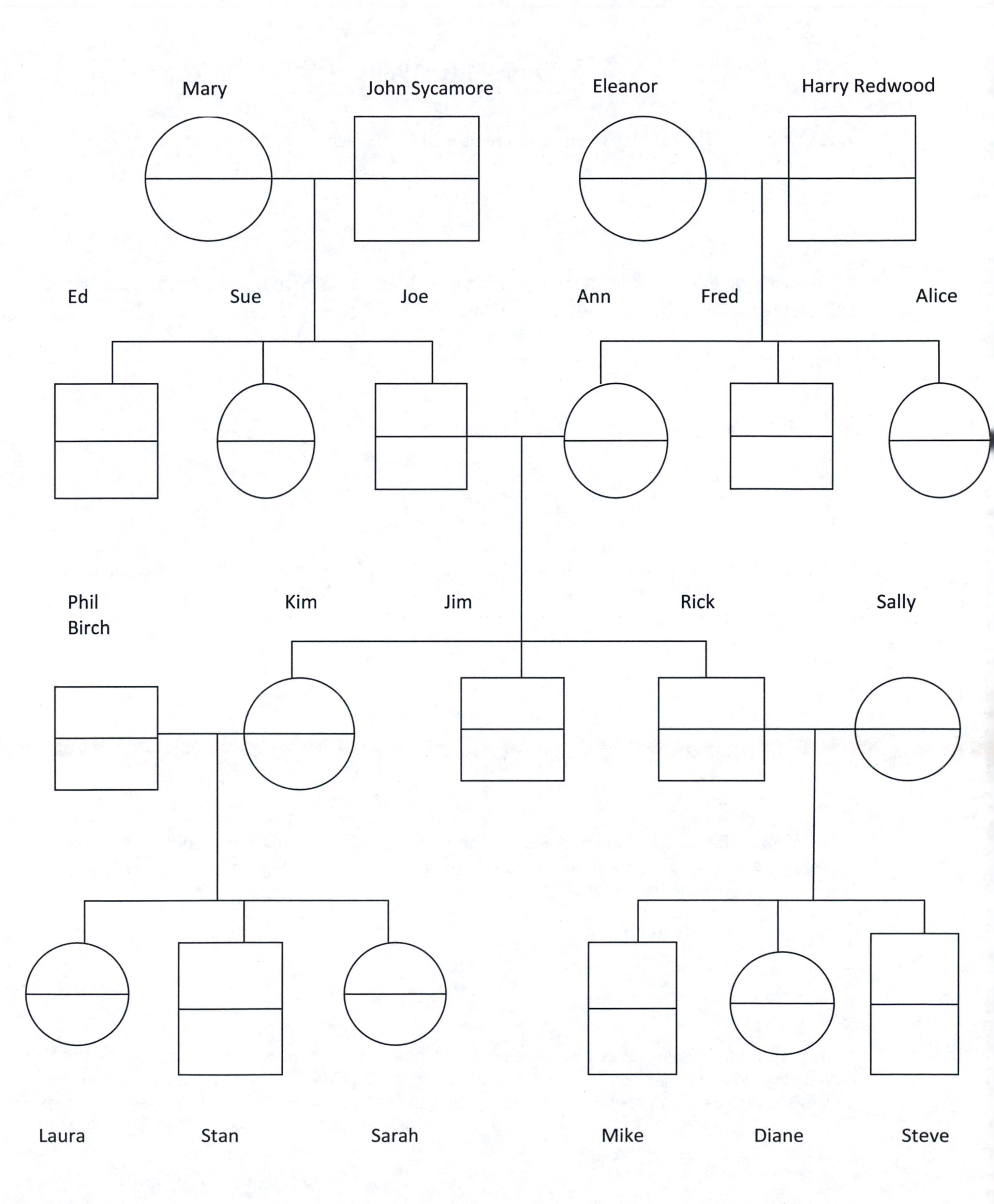
Mary
John Sycamore
Eleanor
Harry Redwood
Ed
Sue
Joe
Ann
Fred
Alice
Phil
Birch
Kim
Jim
Rick
Sally
Laura
Stan
Sarah
Mike
Diane
Steve

After Class Work

1. How many different kinds of gametes are possible with each or the following genotypes?

a. AA b. aa c. Aa

2. In peas, yellow color is dominant to green. Make a key and assign genotypic and phenotypic percentages to each of the following crosses:

a. heterozygous yellow vs. green

b. heterozygous yellow vs. heterozygous yellow

3. White color is dominant to yellow color in squash. A heterozygous white-fruited plant is crossed with a yellow-fruited plant. Make a key and assign genotypic and phenotypic percentages to the cross.

4. In one experiment, Mendel crossed a true-breeding pea plant having green pods with a true-breeding pea plant having yellow pods. All of the F1 plants had green pods. Which trait (green or yellow pods) is recessive? Explain.

5. Among humans, dark eyes (B) is dominant to blue eyes (b). In a family, one parent has dark eyes and the other has blue eyes. Among their offspring, 2 develop dark eyes and 2 develop blue eyes. What are the genotypes of the parents?

6. In a family, one parent has dark hair (D, dominant) and the other has blonde (d, recessive). Among their 9 offspring, all develop dark hair. What are the most likely genotypes for the parents?

7. One gene has alleles A and a; another gene has alleles B and b. For each of the following genotypes, assuming independent assortment what type(s) of gametes will be produced?

a. AABB

c. Aabb

b. AaBB

d. AaBb

8. Using the choices from problem 9, which of the individuals are pure breeding?

9. A man with sickle-cell trait (a mild form of sickle-cell anemia) is married to a woman with the same genotype and phenotype. (Hb = hemoglobin)

Key: $Hb^{A}Hb^{A}$ = normal
$Hb^{A}Hb^{S}$ = sickle-cell trait
$Hb^{S}Hb^{S}$ = sickle-cell anemia

a. What are the chances that this couple will have a child with sickle-cell anemia?

b. Sickle-cell trait?

c. Normal?

10. Thalassemia is a type of human anemia rather common in Mediterranean populations, but relatively rare in other peoples. The disease occurs in two forms, called minor and major; but the later is much more severe. Severely affected individuals are homozygous for the aberrant gene; mildly affected persons are heterozygous. Persons free of this disease are homozygous for the normal allele. For the following problems, let T = the allele for thalassemia and t = its normal alternative.

a. What is the genotype for a severely affected (Thalassemia Major) individual? Mildly affected (Thalassemia Minor)? Normal person?

b. A man with Thalassemia Minor marries a normal woman. With respect to Thalassemia, what types of children and in what proportions, may they expect?

c. Both father and mother in a particular family have Thalassemia Minor. What is the chance that their baby will be severely affected? Mildly affected? Normal?

d. An infant has Thalassemia Major. From the information given so far, what possibilities might you expect to find if you checked the infant's parents for Thalassemia?

e. Thalassemia Major is almost always fatal in childhood. How does this fact modify your answer to 5d?

11. A mother has blood type A and the father has blood type O.

a. What do you know about the genotypes of these parents?

b. The first child of these parents has type A blood. Does this give you any more information about the genotypes of the parents?

c. The second child of these parents has type O blood. What do you know about the genotypes of the parents?

12. A mother has the blood type AB and the father has blood type O. What are the possible blood types of their offspring?

13. The ABO blood system has often been employed to settle cases of disputed paternity. Suppose, as an expert in genetics, you are called to testify in a case where the mother has type A blood, the child has type O blood, and the alleged father has type B blood. How would you respond to the following statements:

a. The attorney of the alleged father: "Since the mother has type A blood, the type O blood of the child must have come from the father, and since my client has type B blood, he obviously could not have fathered this child."

b. The mother's attorney: "Further tests revealed that this man is heterozygous and therefore he must be the father."

14. Both Mrs. Smith and Mrs. Jones had babies the same day in the same hospital. Mrs. Smith took home a baby girl, whom she named Shirley. Mrs. Jones took home a baby girl, whom she name Jane. Mrs. Jones began to suspect, however, that her child had been accidentally switched with the Smith baby in the nursery. Blood tests were made. Mr. Smith was type A; Mrs. Smith was type B; Mr. Jones was type A; Mrs. Jones was type A; Shirley was type O and Jane was type B. Had a mix up occurred?

15. Bar eye in the fruit fly (Drosophila) is an X-linked characteristic in which bar eye is dominant over non-bar eye.

a. Construct a key for bar eyes and non-bar eyes.

b. If a homozygous bar-eyed female is crossed with a non-bar eyed male, what will be the appearance of the offspring?

16. Most types of colorblindness are due to an X-linked recessive allele. A colorblind man and a woman who is heterozygous for the colorblind allele are expecting their first child.

a. What are the chances that a girl born of this mating will be colorblind?

b. What are the chances that a boy born of this mating will be colorblind?

17. A hemophiliac is married to a homozygous normal woman.

a. Will any of their children have hemophilia?

b. If their sons marry a homozygous normal woman, will they have hemophiliac children?

c. If their daughters (from part a) marry normal men, what are the chances the daughters will have hemophiliac girls? Of having hemophiliac sons?

18. Suppose that Gene "b" is sex-linked recessive and lethal. A man marries a woman who is heterozygous for this gene. If the couple had many normal children, what would be the predicted sex ratio of these children?

19. Coat color in cats is determined by a pair of sex-linked genes.

Females: yellow = X^bX^b
black = X^BX^B
calico = X^BX^b

Males: yellow = X^bY
black = X^BY

a. A calico cat has 8 kittens: 1 yellow male, 2 black males, 2 yellow females and 3 calico females. Assuming a single father for the litter, what is the probable color of the father?

b. A black cat has a litter of 7 kittens: 3 black males, 1 black female and 3 calico females. Comment of the probable paternity of the litter.

Answers

1. a. 1—A b. 1—a c. 2—A and a

2. Key: Y = yellow; y = green

a. Yy x yy = 50% Yy 50% yellow
= 50% yy 50% green

b. 25% YY 75% yellow
50% Yy 25% green
25% yy

3. Key: W = white; w = yellow

Ww x ww = 50% Ww 50% white
50% ww 50% yellow

4. Yellow is recessive because the first generation plants must be heterozygous and had a green phenotype, green must be dominant over the recessive yellow. (Green masks out the presence of the yellow trait.)

5. Bb (dark), bb (blue)

6. DD (dark), dd (blonde)

7. a. AB
b. AB and aB
c. Ab and ab
d. Ab, aB, AB and ab

8. only problem 9a is pure breeding—only one type of gamete is produced

9. (male) Hb^AHb^S vs. (female) Hb^AHB^S =

Genotypes		Phenotypes
25% (1/4) Hb^AHb^A	=	normal
50% (1/2) Hb^AHb^S	=	sickle-cell trait
25% (1/4) Hb^SHb^S	=	sickle-cell anemia

10. a. Minor = Tt
Major = TT
Normal = tt

b. Tt vs. tt = ½ Tt
½ tt
Phenotypes = ½ minor, ½ normal

c. Tt vs. Tt = ¼ TT = 25% major
½ Tt = 50% minor
¼ tt = 25% normal

d. To produce a TT there are 2 possibilities:
1. TT vs. TT (major vs. major)
2. Tt vs. TT (minor vs. major)
3. Tv vs. Tt (minor vs. minor)

e. Since majors (TT) die before adulthood, the only possibility is for a minor vs. minor (Tt vs. Tt). Think about the implications of this type or problem for genetic testing and genetic counseling for parents.

11. a. Mother many have I^Ai or I^AI^A. The father must be ii.
b. No new information. The child's genotype, however, must be I^Ai (because the father can only give i and the I^A must therefore come from the mother.)
c. Now you know that the mother must be I^Ai.

12. 50% type A and 50% type B

13. a. The mother must be heterozygous (I^Ai). The man having type B blood could have fathered the child is he were also heterozygous (I^Bi).

b. If the man is heterozygous, then be could be the father. However, because any other type B heterozygous male could be the father, one cannot say that this particular many absolutely must be. Actually, any male who could contribute an i allele could have fathered the child. This would include males with type O blood (ii) or type A blood who are heterozygous (I^Ai).

14. Mr. Smith = type A = I^A__
Mrs. Smith = type B = I^B__
Shirley Smith (?) = type O = ii

Mr. Jones = type A = I^A__
Mrs. Jones = type A = I^A__
Jane Jones (?) = type B = I^B__

A mix-up did occur. Since both Jones are type A, they can only have type A or O children (O if both are I^Ai). The Smiths are types A and B, so they could have type A, B, AB or O children. Shirley could belong to both sets of parents; however, since Jane is type B, she must belong to the Smiths.

15. a. bar eye = X^B non-bar eye = X^b

b. bar eye female = X^BX^B non-bar eye male = X^bY
offspring: males = 100% X^BY, bar eyed
females = 100% X^BX^b, bar eyed

16. The man is X^bY and the female is X^BX^b.
a. For a girl, 50% may be colorblind.
b. For a boy, 50% may be colorblind.

17. Key: Hemophilia = X^h Normal = X^H

a. Hemophiliac Man = X^hY
Normal Homozygous Woman = X^HX^H
Children: All boys are X^HY, normal
All girls are X^HX^h, normal (carriers)
Therefore, 0% will have hemophilia

b. The boys from (a) are X^HY
Homozygous normal woman = X^HX^H
Children: All boys are X^HY
All girls are X^HX^H
Therefore, 0% will have hemophilia

c. The girls from (a) are X^HX^h
Normal man = X^HY
1. Girls will be: ½ are X^HX^H (normal_
½ are X^HX^h (carrier)
All girls are normal.
Boys will be: ½ are X^HY (normal), ½ are X^hY (hemophiliac)

18. Key: Bad gene = X^b Normal gene = X^B

Normal man = X^BY
Heterozygous woman = X^BX^b

Predictions of offspring: X^BX^B = normal female
X^BX^b = normal (carrier) female
X^BY = normal male
X^bY = lethal male, dies

Therefore, since one type of male will die, only 3 possible of surviving offspring are possible: 2/3 are female and 1/3 are male.

19. a. Calico Mother = X^BX^b

Kittens = yellow male: X^bY

yellow female: X^bX^b

black male: X^BY

calico female: X^BX^b

You can't tell anything about the father by looking at the male kittens because they received the Y chromosome (which lacks the color gene) from him. The calico females are also no help because the mother could have given them either gene (X^B or X^b). The clue then is the yellow female. She got one X^b from the mother and the other from the father. Therefore the father is X^bY, yellow.

b. Black Mother = X^BX^B

Kittens = black male: X^BY

black female: X^BX^B

calico female: X^BX^b

Again, the male kitten is of no help (got the Y from the father). The mother can only pass along X^B. Therefore the black female kitten got the other X^B from her father, a black male (X^BY). Yet the calico female kitten had to have gotten the X^b from a yellow male (X^bY). One could assume that the litter was fathered by 2 males: one yellow and one black.

Comment: From these problems, you would be justified in giving good odds in a wager that only females can be calico. You are correct; however, calico males can rarely occur. These cats, however, are sterile and shows there is something abnormal about their sexual development.

Sample Multiple Choice Questions

1. The term "dominant" means that
   a. the dominant phenotype shows up at least 50% of the time in the first generation.
   b. all members of the F_2 generation of a hybrid cross exhibit the dominant phenotype.
   c. one allele can mask the expression of another is a hybrid.
   d. the dominant phenotype shows up in 100% of the offspring in all generations.
   e. the dominant phenotype is more beneficial than the recessive phenotype.

2. Using standard conventions for naming alleles, which of the following pairs is correct?
   a. Tt: recessive phenotype
   b. TT: heterozygous
   c. tt: homozygous
   d. tt: dominant phenotype
   e. all are correct

3. Why is red/green color-blindness more common in males than in females? Because
   a. females would have to receive two copies of the recessive gene to express the trait.
   b. a males only needs to receive the recessive gene from his mother to be color-blind.
   c. color-blindness is an X-linked trait.
   d. all of the above
   e. none of the above

4. Breeding a yellow dog with a brown dog produced puppies with both yellow and brown hairs. This is an example of
   a. dominance.
   b. co-dominance.
   c. incomplete dominance.
   d. epistasis.
   e. polygenic inheritance.

5. If the mother is has type A blood and the father has type AB, what would the blood types of the children be?
   a. all AB
   b. ½ A, ½ B
   c. ½ A, ½ AB
   d. ½ A, ¼ B, ¼ B
   e. insufficient information is available

Answers to Multiple Choice Questions

1. C. A dominant gene masks the expression of a recessive in a heterozygote (hybrid). In a Mendelian cross the F_1 generation is 100% dominant (A) and the F_2 (B) is 75% dominant and 25% recessive. Choice E points out a major misconception about dominant genes. A dominant gene is not always the wild type (dominance) in a population—most prevalent. Also having a dominant gene does not mean having a good gene. The genetic disorder, Huntingtons, is caused by a dominant gene and eventually this mutated, "bad" gene will cause death.

2. C. When both genes are the same (TT or tt) the individual is homozygous. Heterozygous means the genes are different (Tt). In classical Mendelian genetics, T is dominant (will always be expressed) and t is recessive (hidden or masked). If an individual contains at least one dominant gene, they will have the dominant phenotype. The only way to show the recessive phenotype is to have both recessive genes.

3. D. The gene is on the X chromosome. A female has 2 X chromosomes and a male has an X and a Y.

4. B. Each of the genes is expressed. If the pattern was incomplete dominance (C) the puppies would be in the tan shade (a mixture of both colors). If regular dominance (A) was at work the puppies would be either yellow or brown. Epistasis (D) is one gene (resulting protein) is needed by other proteins (genes) to do their job. Polygenic inheritance (E) is many genes that exhibit incomplete dominance to each other to make one trait.

5. E. Since the mother's genotype could be I^AI^A or I^Ai, it is impossible to assign chances to the children. We do know, however, the no matter what genotype the mother is, the children could have the following blood types: B (I^Bi), A (I^AI^A or I^Ai), or AB (I^AI^B).

UNIT 6: BIOTECHNOLOGY

Outline

TOPIC	SOLOMON
I. From Previous Units	
A. Enzymes	
B. Mitosis	
C. DNA	
D. Protein Synthesis	
E. Adaptation vs. Evolution	
F. Mendel	
II. Genetic "Engineering"	
A. Science of the Future	
B. Making Recombinant DNA	314-320
1. Vectors	
2. Restriction Enzymes	
3. DNA Libraries	
4. Genetic Probes	
5. cDNA and reverse transcriptase	291-292
6. PCR	
C. Analyzing DNA	321
1. Gel Electrophoresis	328-329
2. RFLP	
D. Biotechnology Applications	
1. Gene Sequencing	321-325
2. Microarrays	326
3. Products	330-332
4.Gene Therapy	350
5. Cloning	341-343
(a). Stem Cells	359-364
(i.) Types	
(ii). Sources	
6. Ethical Issues	332-333, 365
III. Supplements	
A. Thinkwell CD-ROM	
1. Biotechnology	
a. Plasmids and Gene Cloning	
b. Techniques in Biotechnology	
c. More Techniques in Biotechnology	
IV. Exam	

Objectives

1. Describe the role of vectors in the process of recombinant DNA.
2. Explain the process in making recombinant DNA including cDNA techniques.
3. Explain various laboratory techniques used to amplify DNA. Explain the various techniques used to visualize molecules separated in a gel.
4. Give examples and uses of recombinant engineered products.
5. Define transgenic organisms. Give examples including applications for transgenic plants and animals.
6. What are possible applications of genetic engineering, both present and future, to humans?
7. Define stem cells. Explain the differences between various types of stem cells and give examples for each.

After Class Work

1. After recombinant plasmids have been produced, they are mixed with bacteria to allow the bacteria to take up the plasmids from the solution. Normally, only a small fraction of the bacteria present actually take up the plasmids. Given this information, why do you think that R plasmids, which confer antibiotic resistance, are often used for recombinant DNA work?

2.

_____ A. RFLP Analysis	1. Similar to members of your family but not strangers
_____ B. Paternity	2. Cuts different people's DNA at different points
_____ C. Gel	3. Piece of DNA cut up by restriction enzymes
_____ D. Nucleotide Sequence	4. Place where enzyme cleaves DNA; depends on person
_____ E. Positive Pole	5. Restriction fragment length polymorphism
_____ F. Restriction Fragment	6. Type of cell often used in RFLP analysis
_____ G. DNA polymerase	7. Separates DNA fragments by size and electrical charge
_____ H. RFLP	8. Restriction fragments move through this
_____ I. Genome	9. Restriction fragments are attracted to this
_____ J. White blood cell	10. Where specific restriction fragments collects in gel
_____ K. Band	11. Nucleotide sequence that indicates a specific allele
_____ L. Restriction enzyme	12. Used to locate a particular genetic marker
_____ M. Genetic marker	13. Method for making many copies of a DNA molecule
_____ N. Blood	14. Used to replicate DNA in a test tube for PCR method
_____ O. DNA fingerprint	15. DNA testing can be used to link this with a crime
_____ P. Polymerase chain reaction	16. An individual's electrophoresis band pattern
_____ Q. Electrophoresis	17. DNA fingerprinting is also used to determine this
_____ R. Contamination	18. Only a small part of this is used in forensic DNA tests
_____ S. Restriction site	19. Problem in using PCR in forensic DNA tests

Answers

1. If bacteria are grown in a medium containing the antibiotic, only those bacteria that have actually taken up plasmids will survive. This solves the problem of sorting through bacteria that no not contain recombinant DNA to find the few that do.

2.

A. 12
B. 17
C. 8
D. 1
E. 9
F. 3
G. 14
H. 5
I. 18
J. 6
K. 10
L. 2
M. 11
N. 15
O. 16
P. 13
Q. 7
R. 19
S. 4

Sample Multiple Choice Questions

1. What purpose do restriction enzymes play in bacterial cells? Restriction enzymes
    a. prevent the overproduction of mRNA in the bacterial cell.
    b. attack bacteriophage DNA when it enters the cell.
    c. promote bonding of the promoter to the mRNA molecule.
    d. connect Okazaki fragments.

2. To avoid the introduction on introns into the vector, a _____ copy of mature mRNA is made using the enzyme _____.
    a. sDNA; RNA polymerase
    b. cDNA; DNA ligase
    c. cDNA; reverse transcriptase
    d. sDNA; reverse transcriptase
    e. cDNA; DNA polymerase

3. A _____ is required to transfer genes from one organism to another.
    a. vector
    b. reverse transcriptase
    c. transport molecule
    d. genetic probe
    e. PCR device

4. Which of the following would stop evolution by natural selection from occurring?
    a. if humans became extinct because of a disease epidemic
    b. if a thermonuclear war killed most living organisms and changed the environment drastically
    c. if ozone depletion led to increased ultraviolet radiation, which caused many new mutations
    d. if all individuals in a population were genetically identical, and there was no genetic recombination, sexual reproduction, or mutations.

Answers to Multiple Choice Questions

1. B. A bacteriophage is a virus. The viral DNA needs to use the metabolic processes of the bacteria to replicate. Since the virus does not have an immune system, its restriction enzymes will cut the viral DNA into pieces.

2. C. DNA made from RNA is called cDNA (complementary). Since the DNA will be placed into a bacterial cell the introns must be moved because bacterial cells lack the mechanisms to remove them from RNA. Since the "normal" process of DNA to RNA is reversed, the name of the enzyme is called reverse transcriptase. Ligase joins DNA fragments together. DNA polymerase replicates DNA and RNA polymerase reads DNA and makes RNA.

3. A. The vector used to insert DNA into bacterial cells is a plasmid. Viruses are used for insertion of DNA into eukaryotic cells. Reverse transcriptase uses RNA to make a single stranded piece of DNA. A genetic probe is a small piece of single-stranded DNA with a known sequence of nucleotides and is used to look for complementary sequences in chromosomes. A PCR device is called a thermocycler and is used to facilitate DNA replication in a test tube.

4. D. All of the other choices lead to genetic diversity and thus evolution is possible.

FINAL EXAM

Suggested (individual instructors may modify) topics and point distribution

I. Chemistry (A through E, 9%; F and G, 6%)

- A. Atoms and Isotopes
  - 1. Symbols
  - 2. Composition
    - a. Atomic number and weight
    - b. Electron shells and valence
- B. Oxidation vs. reduction
- C. Bonds
  - 1. Ionic
  - 2. Covalent
    - a. Polar vs. non-polar
  - 3. Hydrogen
- D. Equations—reactant and products
- E. Acids, Bases and Salts
  - 1. pH
  - 2. buffers
- F. Functional Groups
  - 1. Acids and Bases
- G. Carbohydrates, Lipids and Proteins
  - 1. Dehydration synthesis and hydrolysis
  - 2. Composition and types

II. Cell Structure (A and B, 10%; C, 6%)

- A. Organelles
  - 1. Composition
  - 2. Function
- B. Cell Types
  - 1. Prokaryotic and eukaryotic
  - 2. Plant and animal
- C. Cell Membrane
  - 1. Structure and function
  - 2. Transport Systems
    - a. Types of molecules moved by each
    - b. Osmosis problems

III. Physical Science (6%)

- A. Enzymes
  - 1. Composition
  - 2. How they work
  - 3. Environmental effects
- B. Coenzymes
  - 1. Structure
  - 2. Function
- C. ATP
  - 1. Structure and function
  - 2. Phosphorylation Techniques

IV. Energy Transformations

- A. Photosynthesis (7%)
  - 1.Light Dependent Reactions
    - a. Purpose of Photosystems
      - (1) Wavelengths used
    - b. Reactants and products
  - 2. Light Independent Reaction
    - a. Reactants and products for C3 plants
- B. Cellular Respiration (9%)
  - 1. Glycolysis
    - a. Reactants and products
  - 2. Anaerobic
    - a. Types
    - b. Reactants and products
    - c. Purpose
  - 3. Kreb's Cycle
    - a. Reactants and products
    - b. Purpose
  - 4. Electron Transport System
    - a. Reactants and products
    - b. Purpose

V. Cellular Divisions

- A. Nucleotides (A and B, 4%)
  - 1. Composition
- B. DNA
  - 1. Structure
  - 2. Replication

C. Mitosis (5%)
    1. Importance
    2. Chromosomes
        a. Structure
        b. Homologous
        c. Diploid and haploid
    3. M Phase—steps
    4. Plant and. Animal

D. Meiosis (D and E, 5%)
    1. Importance
    2. The Process
    3. Spermatogenesis and oogenesis

E. Meiotic Errors
    1. Non-disjunction—definition

VI. Protein Synthesis (8%)

A. RNA
    1. Structure
    2. Types
        a. Function

B. Process
    1. Transcription
    2. Translation
    3. Mutations

VII. Genetics (10%)

A. Monohybrid Crosses
    1. Terms
        a. Alleles
        b. Genotype vs. phenotype
        c. Dominant vs. recessive
        d. P1, F1, F2
    2. Problems

B. Incomplete Dominance
    1. Problems

C. Multiple Alleles and Codominance
    1. ABO blood type problems

D. Sex Determination
    1. Autosomes and X and Y chromosomes
    2. X linked problems

VIII. Unit 6 (15%)

PART 2:
LAB MANUAL

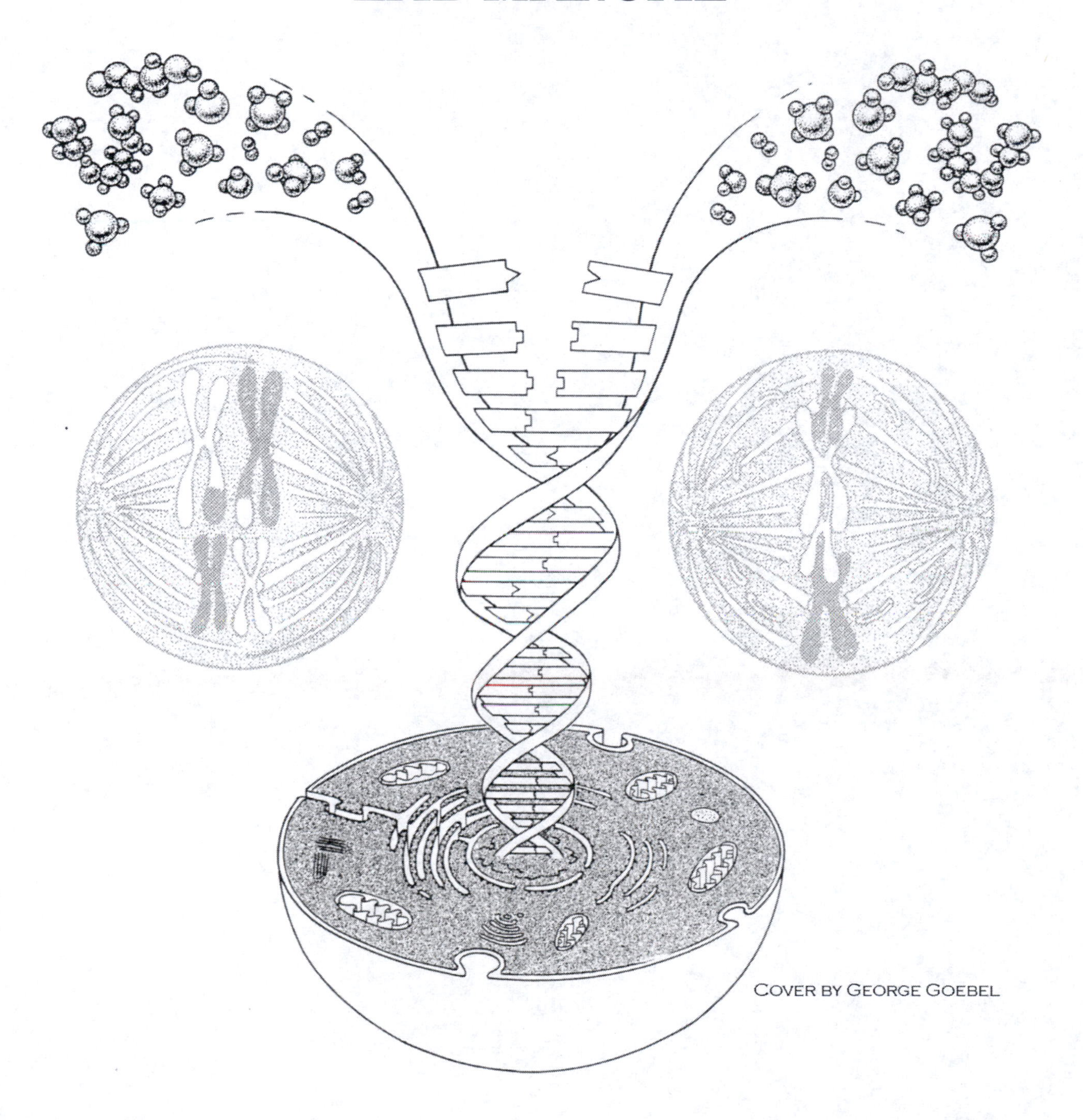

Cover by George Goebel

INTRODUCTION

Welcome to the laboratory component of BIOL 110. You will be spending three hours a week in hands-on activities directly related to your lecture. You will be developing your skills in utilizing the metric system, microscopes, and scientific method as well as having the opportunity to run controlled experiments and interpret data.

Lab Expectations

1. If a class must be missed for <u>any</u> reason, you need to contact your instructor within 24 hours of your class. In the event of a college approved absence, a student is allowed to make up two labs with another instructor (with both instructors permission and there must be space in the lab) and you must get a signed result sheet for validation. If you do not make-up a lab, you will <u>not</u> receive credit for that lab! To receive credit for the lab you must stay until all data has been collected and your station has been cleaned. If a student misses more than 2 labs, they will automatically fail the course.
2. For your own safety, be sure to read lab directions carefully. Make sure you understand what you are doing. Listen carefully for any additional safety precautions. Some general rules that will always hold:
    a. View the Safety Power Point Presentation and pass he safety quiz with an 80%. Also review the Safety Summary that is found in the Appendix. To take part in labs, the safety presentation and quiz must be successfully completed.
    b. You must bring your lab manual and result sheets to class.
    c. You must wear proper clothing:
        i. Closed toe shoes that cover the tops of your feet
        ii. A lab coat that is completely buttoned or pants, with no holes, that come down to your knees; short sleeve shirts and no bare midriffs or backs.
    d. Coats, book bags, purses, etc. are to be stowed under the lab tables or on the coat racks.
    e. Notify your instructor immediately if there are any accidents-broken glass, thermometers, chemicals, etc.
3. Each student is responsible for the maintenance of lab materials. Please return all materials following a lab in the condition -or better than- you received them.
4. At the conclusion of each lab, return your area to its original condition by wiping the surface and washing and drying used materials.

Lab Evaluation

The student's grade will be determined by the following:

- weekly lab quizzes
- informal, short lab reports
- formal lab report

A student must independently pass the lab and lecture portions of BIOL 110 to pass the entire 4 credit course.

EXPERIMENT 1: MEASUREMENT AND THE METRIC SYSTEM

Objectives

1. Make conversions from one unit to another in the Metric System by using the following prefixes: k-, d-, c-, m- and μ-.
2. Utilize appropriate metric base unit for mass, length, volume and temperature.
3. Convert English units of temperature, length, volume and mass to Metric.
4. Properly use the following metric tools: rulers, pipettes graduated cylinder, beakers and/or flasks and balances.
5. Accurately record data on charts to include appropriate units.
6. Properly construct a best fit straight line graph to determine the relationship between the volume and mass of water (density).

Introduction

A Historical Perspective on Measurement

Early man had little need for a system of measurements until he began building things and trading with neighboring tribes. In search of some form of organization, his early units of measurement were based on parts of his body and things he saw around him. Eventually, by using the most convenient objects in his environment, man developed the following units of measurement:

- Cubit: the distance from the point of the elbow to the tip of the middle finger
- Span: the distance from the tip of the thumb to the tip of the little finger with the fingers spread out
- Palm: the breadth of four fingers held together
- Digit: the breadth of the index finger or middle finger

Obviously, problems could occur. For example, if more than one person were doing the measuring for a pyramid or an aqueduct more precise units would be needed. Thus, the measuring rod came into use. The exact length of the rod was usually decided by the local king or pharaoh.

As the power of Egypt declined, the Roman Empire was in ascendancy. The early Romans were the first to measure distance in miles. Mile is short for *mille' passum*, a thousand paces. The term foot was coined by the Greeks, but Romans divided the foot into inches (the breadth of the thumb). These were not exactly the same as the current foot and inch. With the fall of the Roman Empire, their "bureau of standards" was lost.

Again rulers set the standards often using their own body measurements as a guide. King Henry I (1068-1135) defined the yard as the distance from the tip of his nose to the end of his thumb when his arm was fully extended to one side.

As the world became smaller and nations began to communicate with each other, more precise measurements became a necessity. In 1791 the French Academy of Sciences met to discuss the establishment of an internationally accepted system of measurements. This system was the beginning of the metric system. Over the years a series of revisions has taken place, the major ones in 1902, 1913, and 1960. In 1960 the General Conference on Weights and Measures refined the metric system into the *Systeme International d'Unites* (International System of Units), better known as SI. Scientific work requires a quantitative approach. In other words, one that depends on measurement. This system of measurement consists of seven fundamental units (see Table 1.1). It is a modern version of the metric system; units are based on multiples of ten.

Table 1.1: Fundamental Units of Measure in SI

QUANTITY	NAME	SYMBOL
Length	Meter	m
Mass	Kilogram	kg
Volume	Liter	l
Time	Second	s
Electric Current	Ampere	A
Thermodynamic Temperature	Kelvin	K
Amount of a Substance	Mole	mol
Temperature	Celsius	C

The quantities most often measured in the laboratory exercises are mass, length, volume and time. Mass is a measure of the quantity of matter. The SI standard of mass is the kilogram (kg). In lab work the smaller unit gram (g) is more appropriate. The mass of a paperclip is about 1 gram. Length is the distance covered by a line segment connecting two points. The SI unit of length is the meter (m). The meter is about 39 inches. All length dimensions are expressed in multiples of the meter. The modern standard for the meter is based on the wavelength of light emitted by the krypton-86 atom, a sophisticated procedure that can be duplicated precisely. Volume is the space an object or substance occupies. For regular, even-shaped solid objects, it is derived by multiplying length x width x height (V= 1 x h x w), which gives a derived unit of cubic meters (m^3). For laboratory work, the smaller unit cubic centimeters (cm^3), formerly known as a cc, is more appropriate. One cm^3 of water is also equivalent to 1 milliliter (ml) of water. Graduated cylinders are marked in milliliters, so a liquid can be measured directly in milliliters.

Metric Conversions

In the metric system, each unit is a multiple of ten, either greater or smaller than the base unit (see Table 1.2). The most commonly used multiple greater than one is "kilo", meaning a thousand. Adding this prefix to the base unit denotes a 1000-unit multiple: a km, kilometer, represents a thousand meters. The most common multiples denoting quantities less than one are "centi", a hundredth; "milli", a thousandth. "Deci", a tenth and "micro", a millionth are fairly common as well.

Table 1.2: SI Prefixes

Prefix	Symbol	Meaning	Multiplication Factor
tera	T	Trillion	$1{,}000{,}000{,}000{,}000 = 10^{12}$
giga	G	Billion	$1{,}000{,}000{,}000 = 10^{9}$
mega	M	Million	$1{,}000{,}000 = 10^{6}$
kilo	k	Thousand	$1000 = 10^{3}$
hecto	h	Hundred	$100 = 10^{2}$
deka	da	Ten	$10 = 10^{1}$
BASE UNITS(GRAM, METER, LITER, SEC)			
deci	d	Tenth	$0.1 = 10^{-1}$
centi	c	Hundredth	$0.01 = 10^{-2}$
milli	m	Thousandth	$0.001 = 10^{-3}$
micro	μ	Millionth	$0.000\ 001 = 10^{-6}$
nano	n	Billionth	$0.000\ 000\ 001 = 10^{-9}$
pico	p	Trillionth	$0.000\ 000\ 000\ 001 = 10^{-12}$

To make conversions within the metric system, follow the following simple rules:

If you are converting from bigger to smaller units, move the decimal to the right the number of powers of 10 difference between the two units:

1. Convert 1.3 kg to g

- Since kg is bigger than g, the decimal will be moved to the right.
- Since a kg is 1000 g, there are 3 powers of 10 difference between
- kg and g (1000 = 10 x 10 x 10); the decimal is, therefore, moved 3 places to the right.
- 1.3 kg = 1300.0g

2. Convert 142.0 cm to mm

- Since cm is bigger than mm, the decimal will be moved to the right.
- A cm is 0.01 or 10^{-2} m; a mm is 0.001 or 10^{-3} m (there are 10 mm in a cm).
- Therefore, the difference between cm and mm is 1 power of 10 and the decimal point is
- moved 1 place to the right.
- 142.0 cm = 1420.0 mm

If you are converting from smaller to bigger units, move the decimal point to the left the number of powers of 10 difference between the two units:

1. Convert 265.0 ml to l

- Since ml is smaller than l, the decimal point will be moved to the left.
- A ml is 0.001 or 10^{-3} of a l so the difference between ml and l is 3 powers of 10 (0.001 = 0.1 x 0.1 x 0.1). The decimal is moved 3 places to the left.
- 265.0 ml = 0.265 l

2. Convert 7.5 cm to m

- Since cm is smaller than m, the decimal point will be moved to the left.
- A cm is 0.01 or 10^{-2} of a m so the difference between cm and m is 2 powers of 10 (0.01 = 0.1 x 0.1). The decimal is moved 2 places to the left.
- cm = 0.075 m

Common Metric Measuring Tools

Figure 1.1: Linear Measurement—ruler

Figure 1.2: Liquid Measurement—exact measurements

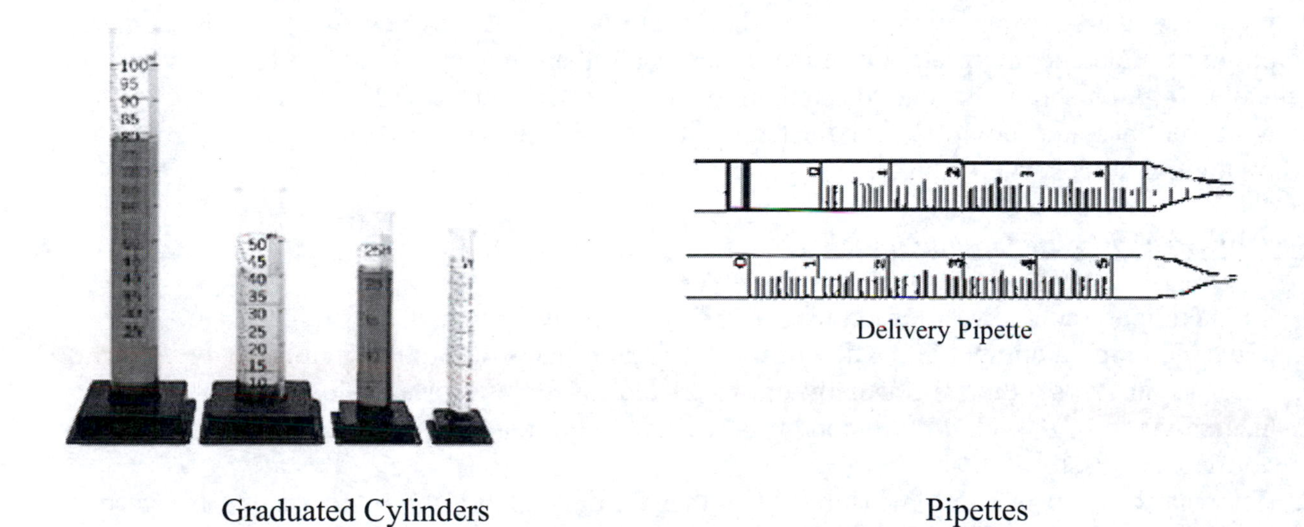

Graduated Cylinders

Pipettes

Figure 1.3: Liquid Measurement—approximate measurements

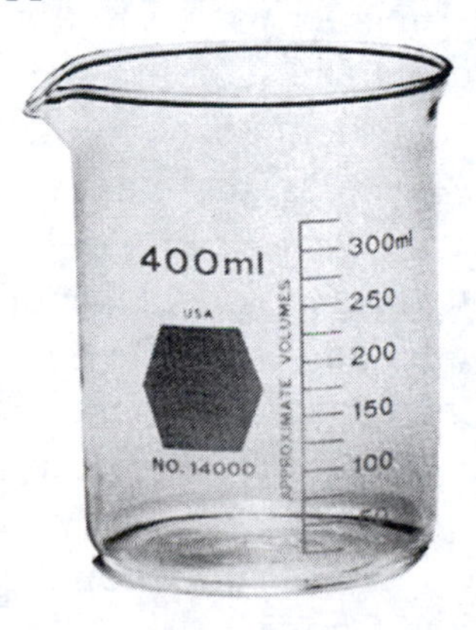

Beaker

Figure 1.4: Mass Measurements

Electronic Balance

Preparing Graphs

A graph is a visual representation of the relationship that exists between two variables. Often a graph gives a clearer interpretation of that relationship than the data alone and thus can be very valuable. A graph should be able to stand on its own; showing the viewer, by proper labeling and drawing of lines, the complete relationship without reference to any other material. Use the following guidelines when constructing graphs.

Best-Fit Straight Line Graph Guidelines

1. In making a graph, always use graph paper, sharp pencils or ink, and a straight edge.
2. Give the graph a number and a descriptive title that reflects the subject matter being graphed.
3. The origin (zero points) is usually drawn in the lower left corner. However, if your data begins far from 0, pick a more appropriate origin. You must use as much of the graph paper for data as possible.
4. The variable that is being measured (hypothetical data) should be plotted along the vertical (y) axis; the variable that is controlled by the experimenter is plotted along the horizontal (x) axis. It doesn't matter if you hold the paper in the portrait or landscape fashion.
    a. For example, if an experimenter wishes to illustrate the relationship between mass and volume of a liquid, he can measure the mass of 3 different volumes of the liquid. Thus, the volume of liquid is controlled by the experimenter and is plotted on the horizontal axis. Whereas, the mass is measured for each volume selected and plotted on the vertical axis. (see Figure 1.5)
5. Label each axis to show the quantity measured and the units used.
6. To determine the distribution of your data, count the number of boxes on the axis then divide the largest value by the number of boxes and round to the nearest whole number that is divisible by 2, 5, 10, etc. Avoid numbers are difficult approximate the space between the lines. Scales may be different for each axis.
    a. For example, you have 3 boxes and the highest value is 125; therefore 125/35 = 3.6 = 5. Each line is worth 5 units.
    b. When numbering the axis label

i. only the darker lines or
ii. every 5th or 10th line.

7. For linear data, plot your data points by finding the x-value on your graph and the y-value. Where they meet, place a point.
8. Draw the best-fit line. In other words, use your straight edge to draw a line on which most of the data points will fall or be very near the line. This line represents an average of all the data that was collected and will be more accurate than just connecting the dots.

Figure 1.5: Example of a two variable graph (such as mass of water vs. volume of water)

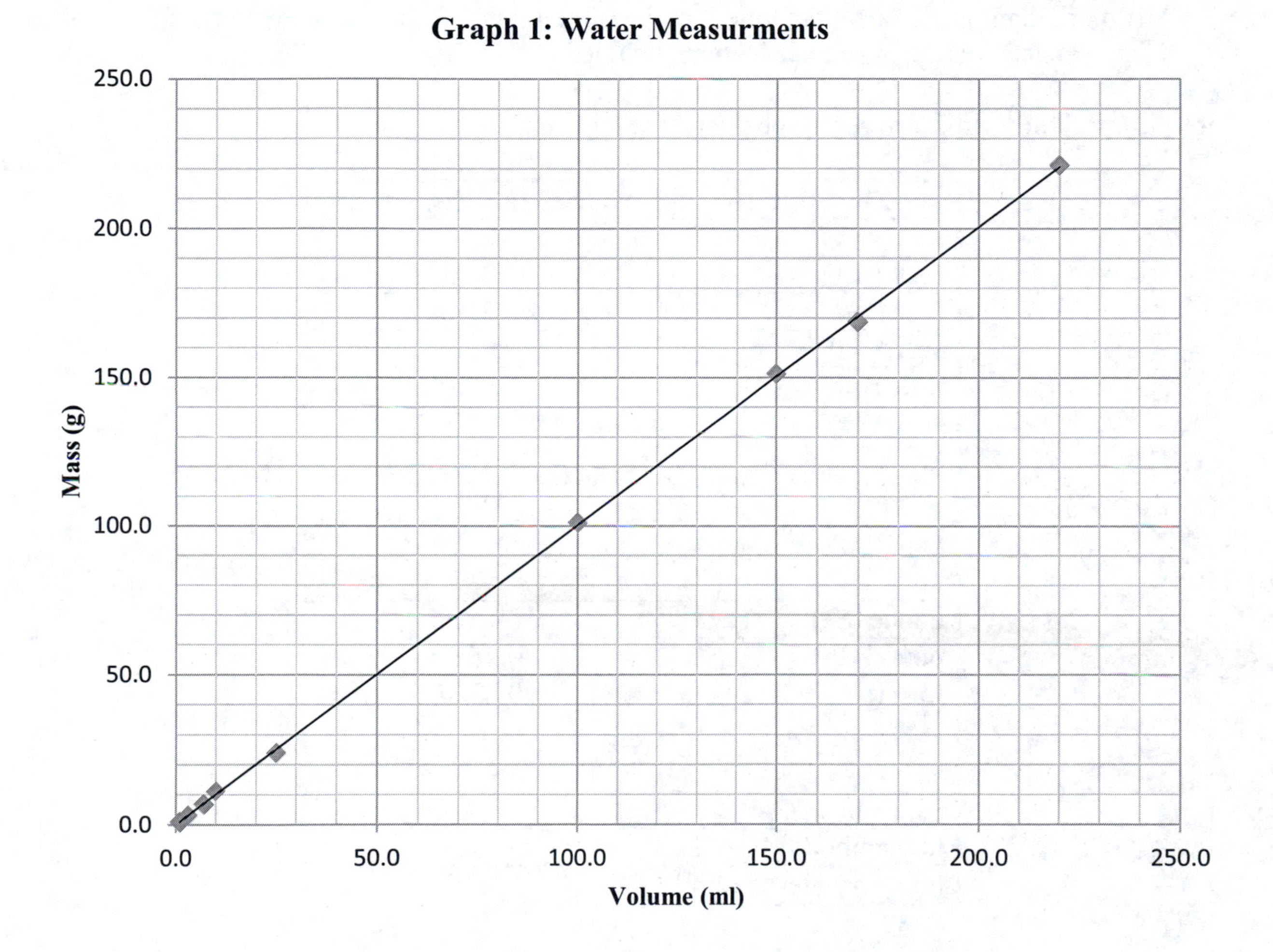

Table 1.3: Requirements for Best Fit Straight Line Graphs

- Proper title
- Axes are labeled with what was measured and the units used
- Determine the range of the data and place the numbers logically and equally along the axes. Do NOT label every line.
- Plot only the data points. Do NOT circle the points nor write the coordinates by the points.
- Use a pen or pencil (crayons and felt tip markers are inappropriate) to construct a best fit straight line.
- If more than one data set is plotted on the same graph, each line needs to be different and a key placed on the graph needs to be included.

Table 1.4: Relationship Between English and Metric Units

English Unit	Metric Unit Equivalent
Linear	
inch (in)	2.54 cm
foot (ft)	0.3 m
yard (yd)	0.9 m
mile (mi)	1.6 km
Liquids	
ounce (oz)	29.6 ml
quart (qt)	0.9 l
gallon (gal)	3.8 l
Mass	
ounce (oz)	28.4 g
pound (lb)	0.45 kg
ton (t)	907.2 kg
Temperature F = 1.8C + 32	
ice forms: 32° F	0° C
human body: 98.6° F	37° C
boiling water: 212° F	100° C

Materials and Methods

Length Measurements

Purpose

1. To use a metric ruler to measure linear dimensions.

Materials

- safety goggles
- metric rulers
- metal rectangle

Procedure

Record all data on the worksheet.

Metric Rulers: All of the numbered (long) lines are 1 cm and the smaller lines are 1 mm.

1. Using the appropriate metric ruler to measure the length of the line below in the following units: centimeters (cm), decimeters (dm), millimeters (mm), meters (m) and micrometers (μm).

12.3 cm

____

2. Using the appropriate metric ruler, measure the length and width of your lab table in meters (m) and centimeters (cm).
3. Measure in cm the length, width, and depth of a metal rectangle. Convert these measurements to dm and mm.

Volume Measurements of a Liquid

Purpose

1. To learn when it is appropriate to use a beaker.
2. To use a graduated cylinder and pipette to measure liquid volumes.

Materials

- safety goggles
- 250 (or 300) ml beaker with ml markings
- graduated cylinders: 50 (or 100) ml and 250 ml
- 1 ml and 10 ml pipettes and pipette helpers
- water

Procedure

Beakers

Even though a beaker has ml markings it is NOT to be used for precise measurements. A beaker is primarily used to prepare solutions or to obtain an approximate volume of liquid.

Graduated Cylinders

To measure the volume of a large amount of liquids, one can simply read the markings on a graduated cylinder or pipette. Choose the graduated cylinder whose maximum volume is closest to the amount of liquid that needs to be measured. The markings or graduations mark off the volume unit in milliliters (ml). When measuring water based liquids, the fluid clings to the sides of the cylinder creating a concave appearance called a meniscus (see Figure 1.6). For best accuracy, always measure at the bottom of the meniscus.

Figure 1.6: Water and water-based liquids form a meniscus

Always measure at the bottom of the meniscus.

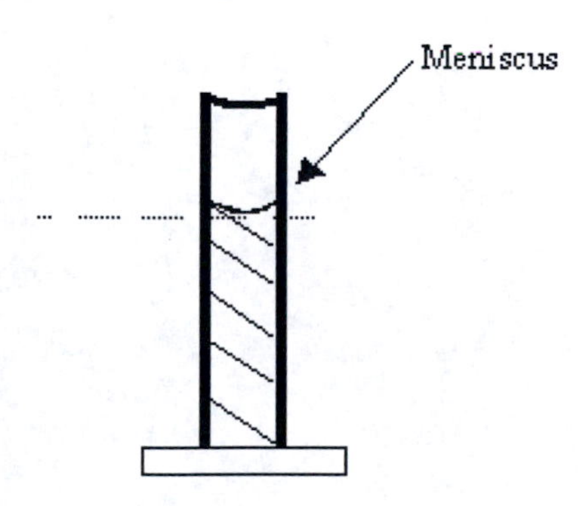

Pipettes

Pipettes are usually used to measure volume of 10 ml or less. With a delivery pipette (Figure 1.7) the 0 is at the top of the pipette. To use this pipette, fill to the 0 line and then "deliver" the required amount (by "lowering" the liquid to the desired amount) to the appropriate container. Mouth pipetting is NEVER permitted. To draw liquid into the pipette a pipette helper is attached. In this class, the green pipette helper is used for the 5.0 and 10.0 ml pipettes and the blue pipette helper is for a 1.0 ml pipette.

<u>Figure 1.7: Delivery Pipette</u>

7.0 ml of liquid were placed into a container; 3.0 ml remain in the pipette

1. Using the markings on the beaker, add 50.0 ml of water to the beaker.
2. Pour this water into the appropriate graduated cylinder. Record the volume.
3. Discard the water.
4. Repeat steps 1-3 but this time use 100.0ml of water.
5. Repeat steps 1-3 but this time use 150.0 ml of water.
6. Repeat steps 1-3 but this time use 200.0 ml of water.
7. Fill a 1 ml pipette with water. Dispense a 0.7 ml. Record the amount of water left.
8. Repeat step 7 but dispense 0.4 ml of water.
9. Fill a 10 ml pipette. Dispense 7 ml. Record the amount of water left
10. Repeat step 9 but dispense 3.5 ml of water.
    NOTE: Never mix used pipettes with clean ones.

Mass Measurements

Purpose

1. To use an electronic balance

Materials

- safety goggles
- electronic balance
- 1 ml and 10 ml pipettes
- 100 ml and 250 ml graduated cylinders
- weigh boat
- 300 ml beaker

Mass is measured by using a balance and is done in grams. To protect the balance pan and make measurement of mass more convenient, a piece of paper (weighing paper), a plastic dish (weighing boat) or another container is used to hold the object being measured. The mass of the container must be subtracted from the total mass to obtain the mass of the material itself. Electronic balances allow you to zero the balance so that the mass of the container is not contained in the readings.

Procedure

1. Turn on the balance by pressing the On/Off button.
2. If using a holding container, tare the balance by placing the container on the pan. Wait for the mass to show in the display then lightly tap the On/Off button and the display will now read 0.0.
3. Add 1.0 ml of water to the weigh boat. Record this data. Determine the mass of the water.
4. Empty the weigh boat and completely dry the weigh boat.
5. Repeat steps 2-3 to determine the mass of 3.0 ml, 7.0 ml and 10.0 ml of water.
6. Using the beaker as the holding container, re-tare the balance. Determine the mass of 25.0 ml, 100.0 ml, 150.0 ml, 170.0 ml and 220.0 ml of water. (Note: Measure the water using the graduated cylinder and then pour the water into the beaker.)

Results

How to Write Numbers

A digit must be placed before and after a decimal point. When a column header does not contain units, you need to add the units to the header or beside each measurement.

Wrong	Right
10 cm	10.0 cm
.7 ml	0.7 ml

Length Measurements (Note: units are included in the column headers)

Line in Manual	cm	dm	mm	m	µm
	12.3	1.23	123	0.123	1230.000

Lab Bench Measurements	m	cm
Length	1.625	162.5
Width	1.215	121.5

Metal Rectangle	cm	dm	mm
Length	7.5	.75	75
Width	2.5	.25	25
Depth	0.6	.06	6.0

Volume Measurements (Note: You need to add units to these tables.)

Beaker and Graduated Cylinder Volumes

Beaker Volume	Graduated Cylinder Volume

Using a pipette

Volume Remaining	Volume Discarded
9.3mL	0.7
0.6 mL	0.4
3.0 mL	7.0
6.5mL	3.5

Mass Measurements

Volume of H_2O	Mass of H_2O
10mL	.9g
3.0	2.9
7.0	6.8
10.0	9.8
25.0	23.8
100.0	97.1
150.0	153.8
170.0	173.1
220.0	223.5

Conclusions

1. Perform the following conversions.
   - a. 0.905 kg to g
   - b. 0.307 mg to g
   - c. 822.0 ml to l
   - d. 0.075 g to mg
   - e. 0.667 mg to cg
   - f. 0.888 kg to g
   - g. 0.0446 km to cm
   - h. 57.5 cg to mg
   - i. 3.35 μl to ml
   - j. 13.5 dl to cl

2. You need to purchase 0.5 kg of cement for you home improvement project. The store only stocks 100 g bags. How many bags will you need to purchase? Show your work.

3. Which is greater (larger)?
   - a. 1 m or 1 yd
   - b. 1 l or 1 qt
   - c. 1 lb or 1 kg
   - d. 1 oz or 1 g
   - e. 1 m or 1 km
   - f. 1 cm or 1 in
   - g. 1 μg or 1 cg
   - h. 1 kg or 1 t
   - i. 1 ft or 1 m
   - j. 1 cl or 1 ml
   - k. 1 kg or 1 g
   - l. 32° F or 32° C

4. Using the Mass Measurements data construct a best fit straight line graph. For graphing help go to: http://faculty.ccbcmd.edu/~kdalton. On the left side on the page, under Tutorial Help, click on Graphing.
   a. By determining the slope of your line you will be able to determine the relationship between the volume of water and its mass (density). To calculate the slope, the following data must be obtained from the graph.

	1st Coordinates	2nd Coordinates
y-axis, g		
x-axis, ml	50	180

Complete the above table and then use the following formula to calculate the slope (m) of your line.

$$m = \Delta y/\Delta x \qquad \text{therefore } m = (y_2 - y_1)/\,(x_2 - x_1)$$

Show your calculation and round your answer to the nearest whole number.

   b. What is the specific relationship between the volume of water and its mass? (Density of water)

5. Using your graph determine the following:
   a. What is the mass of:
      (i) 24.0 ml of water? ii. 193.0 ml of water?
   b. What is the volume of
      (i) 8.0 g of water? ii. 168.0 g of water?

EXPERIMENT 2:
USE OF THE SCIENTIFIC METHOD

Objectives

1. Describe the basic steps of the Scientific Method.
2. Apply the Scientific Method to solve a given problem.
3. Differentiate between negative and positive controls.
4. Describe and contrast science, pseduo-science, bad science, and scientific misconduct.
5. Conduct an experiment to determine if all sugars are equally fermented yeast.

Introduction

(See Chapter 1 in textbook)

Scientific Method is the general approach scientists use to explain occurrences in the natural world by way of systematic observation coupled with systematic testing. The scientific method usually begins with the observation of some occurrence which results in the statement of a problem (a question one would like to answer). After making the observation and stating the problem, inductive reasoning (reasoning proceeding from specific detail to a general statement) is used to formulate a hypothesis. A hypothesis is a tentative explanation that can be tested. It is an educated answer to your question and is based on current knowledge.

After stating the hypothesis, deductive reasoning (begins with a general statement and infers a specific conclusion) is used to determine the types of experiments or observations that will be necessary to support or refute the hypothesis. In other words, one designs experiments which will test the hypothesis. The variable in the experiment is also called the independent variable and your expected results are the dependent variable.

The experiments allow the experimenter to collect data (all the evidence which is collected, whether it supports or refutes the hypothesis). To be valid, data must come from experiments that are repeatable and were designed with appropriate controls (samples going through all steps of the experiment except the one being tested). Other types of controls may also be used that will determine if required reagents are properly working. These types of controls are called positive (where the sample contains a known sample that will turn the reagent the required color) or negative (where the sample, usually water, will not change the color of the reagent). From the results of the experiments, conclusions are drawn. However, science is very cautious. Scientific observations are never totally complete. Therefore, a hypothesis can never be proven true; it can only be supported or refuted by the collected data. When a hypothesis is supported by a large number of observations and is considered valid by a vast majority of scientists, it is considered a scientific theory.

This lab provides two unique opportunities to use the scientific method. The first activity is a case study which provides a brief, factual presentation of the work of Ignaz Semmelweis in his efforts to remedy the problem of childbed fever that was frequently seen in Europe in the mid-1800's. The second activity involves metabolism of sugars by yeast.

An Understanding of the Scientific Method—Dr. Dave O'Neill

The scientific method is an approach used to test ideas about reality based observations, testing assumptions, developing evidence, applying logic, and drawing conclusions. Important ideas that have been tested and not found to be wrong are more likely to be true. Ideas that can be tested, and if wrong, are called falsifiable ideas. Falsifiable ideas that have been tested are often more useful to scientists, engineers, and technicians than ideas that are not testable or have not been tested.

The method is not fixed but often includes observations of phenomena, development of a hypothesis (a formal answer to a proposed question) that explains the observation. Once developed, the hypothesis is either rejected or supported. Experimentation that is designed to test a hypothesis must be repeatable; therefore, the material and methods of the test are defined and recorded and standard units of measurement are used. Results of the tests and the logic used in drawing conclusions are reported or published in order to allow review and criticism of all facets of the process by other interested in the topic discussed.

After a hypothesis is tested, reviewed, and not rejected it is tentatively accepted as correct, with the understanding that the idea may be rejected if further evidence contracts it. When the idea explains some aspect of the natural world and can incorporate facts, laws, inferences and tested hypothesis, this important idea is then called a theory. If a theory has wide-ranging significance and stands up to many tests by independent investigators, that theory may be called a principle. Principles are sometimes thought as "truths" although scientists tend to avoid that word along with proof and fact.

Controlled experiments (that reduce the likelihood that the influence of some untested factor alters the results) and statistical tests are applied. Often data is presented as graphs that show the mathematical relationship between two factors.

Blind experiments are designed to reduce the effect of bias on an outcome. Bias often results from strongly held convictions or beliefs called paradigms. It is essential that everyone interested in the correctness of the idea be aware of their paradigms and makes every effect to avoid developing conclusions based on paradigms rather than evidence.

Science is a process that is used to develop and test ideas about the natural universe. The scientific process has resulted in a body of knowledge that is supported by evidence. That body of knowledge is called "science". Science has led to the idea that the existence of the universe and events that occur in the universe are the result of natural forces and that the behavior of those forces can be understood. Study of the way things work has led to the recognition of certain relationships called Principles that seem to apply throughout the universe. The idea that the universe works according to these principles is called a mechanistic approach to understanding the universe and is typical of scientists. A spiritual approach to the understanding of the universe explains events in terms of supernatural forces (spirits, energy, forces or God(s)). It is important to understand that the existence or characteristics of these entities may not be demonstrable by scientific methods.

Attempts to make spiritual ideas or beliefs that do not survive experimental testing but appear to be scientific are called pseudo-science. Poor experimental design that allows false ideas to survive is called bad science. If experimentation or results are falsified to achieve a particular outcome the process is called scientific misconduct. If pseudo-science, bad science, or scientific misconduct is deliberately used to obtain money, the action may be a criminal act of fraud.

Science is the process of developing and testing ideas. Engineering puts scientific findings to practical use. Technology makes the applications available to large numbers of people. In a sense, the existence of technology validates the scientific method.

Spiritual belief systems, such as religions, are usually not based on science. The core beliefs of religions are generally not falsifiable. Belief is a powerful motivator of human behavior. Religions, for example, have been the motivating force for holy wars (crusades and jihads) that have resulted in untold human suffering. On the other hand, religions provide comfort in times of great human sorrow and motivate charitable acts that relieve human suffering. Likewise, the applications of science (engineering and technology) can be used for evil (eugenics, weaponry) or good (medicine, medical technology, transportation, etc.). Many scientists are religious and seek to find the positive aspects of both science and religion in their lives. Others are not religious.

photons- packets of energy

Materials and Methods

Part A: A Case Study

Adapted from "Case Studies in Science: State University of New York at Buffalo"

Materials

- Case study cards

Procedure

The case study contains four parts; each followed by Study Questions. After reading the evidence, you will be expected to discuss the questions with your partner and then share your discussion with the entire class. Evidence from one part will be needed for future parts.

Part B: Sugar Metabolism by Yeast

Adapted from "Vernier: Biology with Computers"

Yeast are able to catabolize (a metabolic reaction that breaks down a macromolecule to release the energy stored in the covalent bonds) certain molecules. In order for an organism to make use of this potential energy, the organism must have the proper enzymes that will break these bonds. These bonds can be broken with the aid of oxygen (aerobically) or without oxygen (anaerobically).

The general reaction for aerobic metabolism is:

$$\text{sugar} + O_2 \rightarrow H_2O + CO_2 + \text{released energy}$$

The general reaction for fermentation is:

$$\text{sugar} \rightarrow \text{ethanol} + CO_2 + \text{released energy}$$

When yeast use aerobic processes, oxygen consumption is occurring at the same rate as carbon dioxide production. If we monitor gas pressure, this reaction will not cause a pressure change in a closed environment. Since fermentation only involves one gas, we will be able to monitor the reaction by observing an increase in CO_2 which will cause an increase in gas pressure.

Materials

- safety goggles
- Windows PC with Vernier software
- 10 ml pipettes with pipette helpers
- small beaker
- 18 x 150 mm test tube
- test tube rack
- Vernier Gas Pressure Sensor
- Vernier Computer Interface Logger *Pro*
- heating block set to 38°C
- 1-hole rubber stopper assembly
- plastic tubing with Luer-lock fitting
- yeast suspension (3 g yeast/75 ml water)
- 1 of the following:
    - 5% glucose
    - 5% sucrose
    - 5% lactose
    - 5% fructose
    - water
- vegetable or mineral oil with dropper pipette

Procedure

*Steps to be done by the instructor.

1. *Open Logger Pro: if the bottom part of the screen is not gray/empty then close all the windows by clicking on the appropriate X.
2. *Prepare the computer for data collection by opening Experiment 12B: Gas Pressure, MBL file from the *Biology with Computers* folder of Logger*Pro*. (File-open) The vertical axis has pressure scaled from 90 to 200 kPa. The horizontal axis has times scaled from 0 to 15 minutes. The data rate is set to 6 samples/minute.
3. *Make sure that the plastic tubing from the rubber stopper in connected to the Gas Pressure Sensor which is connected CH 1 on the Vernier LabPro Box. On the bench top will also be a tan cable, this is presently connected to the computer and the LabPro. Make sure the LabPro is connected to the wall outlet.
4. Place 2.5 ml of your solution into a test tube.
5. Place the tube into the heating block for 5 minutes.
6. Add 2.5 ml of yeast. Gently swirl to mix in the yeast.
7. Add 1 dropper full of vegetable oil (the surface needs to be completely covered with the oil). Do not get oil on the inside wall of the test tube.
8. Insert the stopper into the tube. Firmly twist the stopper for an airtight fit. Return the test tube to the heat block.
9. Incubate for 10 minutes.

10. Click the *Collect* icon on the computer to begin to collect the data. Collect data for 15 minutes.
11. Monitor the pressure readings displayed in the window.
12. The computer will stop collecting data after 15 minutes.
13. Click on the Linear Regression (Icon is R=) button. A best-fit straight line will be shown. The line from the box points to the graphed data. In Table 1, record the value of the slope (kPa/sec), m.
14. Discard the contents of the tube and thoroughly wash all glassware. Store the test tube upside down in the test tube rack.
15. Obtain the slopes for the other sugars.
16. Construct a bar graph of your data.

Bar Graph Guidelines

Bar graphs can be used to compare categories. In this experiment you will be comparing the fermentation data obtained from the four different sugars and water.

1. Properly title the graph.
2. The X axis contains the grouped data, in this case the different sugars used. Spread out the groups so that the data bars will not be touching each other.
3. The Y axis contains the frequency (collected) data, in this case the fermentation rates. Construct this axis as you did for a best fit straight line graph.
4. Using a ruler, construct the bars. The width of the bar should be at least 2 squares.

Results

These questions are for the fermentation experiment.

1. Record at least five observations. (Remember that observations are what you know <u>before</u> you do the experiment—appropriate scientific background information.)

2. What is the question (purpose) of your experiment?

3. What is your hypothesis?

Table 1: Fermentation Rates

Contents of Tube	Fermentation Rate (m = kPa/sec)

Conclusions

The following questions refer to the fermentation experiment.

1. List 5 conditions that were controlled in your experiment.
2. What condition was varied in your experiment?
3. What is the purpose of the oil layer?
4. What is the purpose of the yeast + water trial?
5. Which sugar is best utilized by the yeast for fermentation? The worst? Explain by using your data in your response.
6. Why do you need to incubate the yeast before you start monitoring air pressure?

EXPERIMENT 3:
THE SPECTROPHOTOMETER

Objectives

1. Explain the difference between a quantitative and qualitative analysis.
2. Describe the relationship between absorbance and transmittance.
3. List the steps in the proper use of a spectrophotometer.
4. Use an absorbance curve to determine the maximum absorbance.
5. Be able to prepare a blank.
6. Prepare a standard curve and use it to determine the concentration of a solute in a solution.

Introduction

The energy of the sun (electromagnetic spectrum) travels in discrete energy packets called photons which travel at different wavelengths (λ and are measured in nm, 10^{-9} meter). The shorter the wavelength the more energy the light contains. The entire electromagnetic spectrum (from long λ to short λ) is comprised of radio waves, microwaves, infrared, visible light, ultraviolet, X-rays and gamma rays. Wavelengths of 700 to 400 nm comprise the visible light spectrum (red, orange, yellow, green, blue, indigo, violet) and contain enough energy to excite photoreceptors within the human eye so that we can see objects.

Table 3.1: Wavelengths of Visible Light

COLOR	WAVELENGTH (nm)
red	630-700
orange	590-630
yellow	560-590
green	490-560
blue	450-490
indigo	420-450
violet	400-420

Molecules either absorb or transmit energy. The color of an object is determined by how much visible light energy is absorbed or transmitted. Some of the light is absorbed and the rest is reflected. These transmitted wavelengths are the colors that our eyes see. A spectrophotometer is an instrument designed to detect the amount of radiant energy absorbed and transmitted by molecules in a solution. When a specific wavelength of light is aimed at a solution, some of the light energy is absorbed (absorbance, A) and the rest is transmitted through the solution. Absorbance is dependent on the ability of the molecules dissolved (solute) in the solvent to absorb the wavelength of light, the concentration of the solute and the distance of the light from the solution.

The Spectophotometer (Figure 3.1) is capable of generating wavelengths of light from 375 to 625 nm. The selected light (8) is then passed through the sample (2) and is picked up by a measuring phototube where the light energy is converted to an electrical current that is read on the meter (3). The Spec20D has two scales: percent transmittance, which runs from 0 to 100, and absorbance, which runs from 0 to 2.

When a colored solution is then placed into the Spectrophotometer, some of the light will be absorbed by the solute particles and some will be transmitted to the phototube. The amount absorbed will be proportional to the concentration (number of dye molecules per unit volume) of the solution. You will be using absorbance units because they are directly and linearly related to the concentration of the dye in the solution.

Figure 3.1: Spectrophotometer

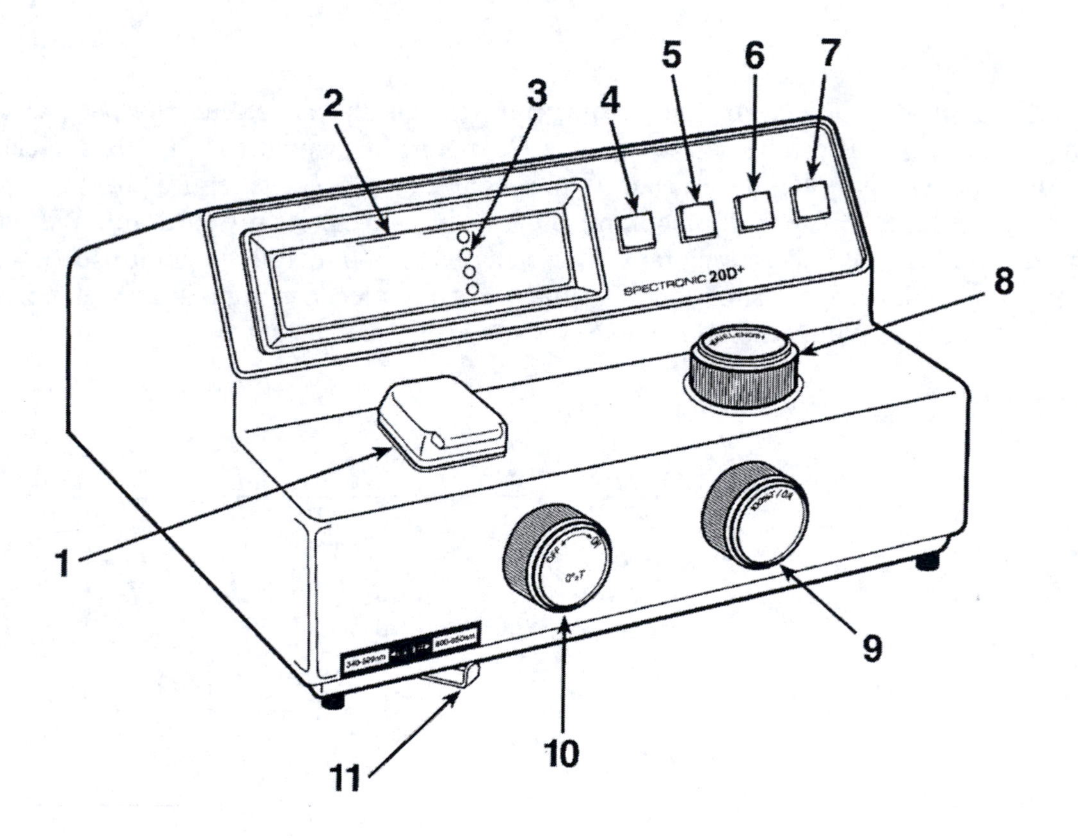

Before the Spectrophotometer can be used to measure the light absorbing properties of a solution, the machine must be calibrated by making three adjustments.

1. The correct wavelength of light (8) must be chosen. This will be the color of light that is best absorbed by the compound under investigation. The filter (11) must be put in place for the corresponding wavelengths:

344-599 nm	Lever to the left
600-950 nm	Lever to the right

2. The output of the phototube must be adjusted to correct for a drift in the electric current. This is accomplished by turning left knob (10), with no sample in the instrument, so that the meter reads 0% transmittance.

3. Compensation must be made for all material other than the solute that is between the light source and the detector. A tube containing only the solvent (the blank) is placed into the instrument and the right knob (9) is adjusted so the meter reads 0 absorbance.

<u>Your Experiment</u>

There are two basic types of analyses that can be done to determine the concentration of bromophenol blue in an unknown sample.

Qualitative Analysis: Identify the components of a substance in a mixture: is bromophenol blue present or absent in the sample.

Quantitative Analysis: Determine the absolute or relative concentration of one or several particles in a substance. By using the spectrophotometer, students will determine the exact concentration of bromophenol blue in an unknown sample.

Materials and Methods

Determine Maximum Absorption for Bromophenol Blue

Materials (for all parts)

- safety goggles
- gloves
- Spec 20
- curvettes
- 10 ml pipettes
- beaker with water
- stock solution of bromophenol blue
- unknown sample of bromophenol blue

Procedure

* Steps to be done by the instructor.

1. * Turn the power on (10) and allow the instrument to warm up for at least 15 minutes.
2. * Press the Mode Button (4) so that the light indicating Transmittance (T) is illuminated.
3. * With NO tube placed into the instrument (1), adjust the meter to 0% T by using the Power Switch—left knob (10). Press the Mode Button (4) so that the light indicating Absorbance is illuminated.
4. Obtain two clean tubes. Make a blank by adding 8 ml of distilled water (solvent) to one tube. Add 8.0 ml of bromophenol stock solution (0.015 mg/ml) to the other tube.
5. Set the wavelength to 425 nm and move the filter lever to the left (11).
6. Using a Kim Wipe™, clean the outside of the blank and insert the tube into the instrument. Do not place your fingers on the tube. Close the cover and adjust the Reference Dial—right knob (9) so that the meter reads 0 absorbance.
7. Remove the blank, close the door.
8. Clean the bromophenol blue tube and insert it into the instrument.
9. Close the door and read the absorbance. Record the reading (2 decimals).
10. Remove the sample. You will repeat steps 5 through 9 for each of the following wavelengths: 450, 475, 500, 525, 550, 575, 600 (Move the filter lever to the right.) and 625 nm. Remember to recalibrate (steps 6 and 7) for each new wavelength.
11. Graph wavelength versus absorbance. Using the graph, determine the wavelength of maximum absorbance for bromophenol blue.
12. The blank will be Tube 1 and the bromophenol blue tube will be Tube 9 in the next part.

Standard Curve

Here you will construct a graph that will show the relationship between the concentration of solutes in a solution and its absorbance. Since you are working with precise concentrations of bromophenol blue this is a quantitative analysis. If you were just interested in whether bromophenol blue was present or absent in the solution you would perform a qualitative analysis.

Procedure

1. Using the stock bromophenol blue solution (0.015 mg/ml) and distilled water prepare 9 tubes as follows. Do NOT write on the tubes. Keep them in order in your test tube rack! Using a gloved hand, cover the top of the tubes and invert three times to mix the solution.

Tube Number	ml Dye	ml H_2O	Concentration
1	0.0	8.0	0
2	1.0	7.0	.001875
3	2.0	6.0	.00375
4	3.0	5.0	.005625
5	4.0	4.0	.0075
6	5.0	3.0	.009375
7	6.0	2.0	.01125
8	7.0	1.0	.013125
9	8.0	0.0	.015

2. Determine the concentration of bromophenol blue in each of the tubes by using the following formula.

   (ml of dye x concentration of stock)/total volume in tube = mass of dye/ml)

   therefore for our problem (ml of dye x 0.015 mg/ml)/8.0 ml = mg dye/ml)

   Record this number in the above chart.
3. Set the Spectrophotometer to the maximum absorption wavelength (and filter) as determined in Part 1.
4. After calibrating the instrument with Tube 1 (blank) read the absorbance for all eight tubes in the dilutions series.
5. Graph the concentration of the dye versus the absorbance. (The y-axis needs to include the wavelength that was used.) This graph is called a standard curve. By reading the graph you can determine the concentration of an unknown bromophenol blue solution by knowing the absorbance of the solution at a certain wavelength.
6. Clean the tubes and place them upside down in the test tube rack.

Determination of an Unknown

Procedure

1. Place 8.0 ml of your assigned unknown into a clean tube.
2. Using the same wavelength (and filter) as you used in Part 2, determine the absorbance of this solution.
3. From your graph, determine the concentration of the bromophenol blue that is in your unknown.
4. Clean the tube as before.

Results

1. Determine Maximum Absorption for Bromophenol Blue

Insert Scattered

Wavelength	Absorbance Units
425	.28
450	.31
475	.27
500	.26
525	.33
550	.58
575	1.90
600	1.13
625	.34

a. What is the wavelength of maximum absorbance for bromophenol blue?
b. What color is this?

2. Standard Curve for Bromophenol Blue

Tube Number	Concentration	Absorbance Units
1		0
2		.185
3		.380
4		.576
5		.756
6		.960
7		1.1120
8		1.270
9	.015	.965

3. Determination of an Unknown

Unknown Letter: C

Absorbance Value: 1.27

Conclusions

1. Describe the relationship between absorbance and the concentration of a light-absorbing substance in a solution. based on graph

2. Explain how standard curves can be used to determine the concentration of a substance in a solution. line of best fit linear line

3. You have a solution that appears green. What color light is/are being transmitted? Absorbed? green / all colors except green

4. What is the purpose of a blank when using a spectrophotometer? to zero it out, have a control

5. You are making a standard to be used to measure protein concentration. You have added 2.0 ml of protein, 3.0 ml of water, and 4.0 ml of an indicator dye. How would you prepare the blank for this experiment?

6. To a test tube you add 3.0 ml of a stock 0.05 mg/ml dye and 7.0 ml of water. What is the final concentration of dye in this tube? (Show your work.)

7. What is the concentration of bromophenol blue in your unknown sample?

9mL

(3.0 x 0.05 g/mg / 7mL

=.0214

EXPERIMENT 4:
MICROSCOPES

Objectives

1. Identify the components of the compound light microscope and state the functions of those parts.
2. Define the following and be able to use them correctly in reference to a microscope: resolution, magnification, contrast, parafocal, ocular scale, depth of field and field of view.
3. Demonstrate the proper handling, use and storage light microscope.
4. Using a light microscope, demonstrate the ability to properly focus a slide under scanning, low and high powers.
5. Calculate the total magnification when given ocular and objective powers.
6. Using an ocular scale, determine the size of a microscopic object.
7. Indicate the major advantages & disadvantages of the following types of microscopes (See Chapter 4 in the textbook.):
    a. brightfield
    b. transmission electron
    c. scanning electron

Introduction

Microscopes

Because biological objects can be very small, it is often necessary to use a microscope to view them. Many types of microscopes are available today, including the phase-contrast, interference, dark field, reflecting, polarizing, X-ray, fluorescent, UV, and electron microscopes. We will use the conventional compound light microscope.

The compound light microscope is used for examining small or thinly sliced cross sections or longitudinal sections of objects. Illumination is from below, and the light passes through clear specimen sections, but not through opaque specimens. To improve contrast, stains or dyes that absorb light can be used.

Microscopy can be fascinating or frustrating. Success in using the microscope depends on how well you understand the mechanics of the microscope and the ability to follow proper focusing procedures. With patience you can discover an exciting world with your microscope.

The ability to view these various cell organelles has been greatly enhanced by the invention of the transmission electron microscope. In this type of microscope, electrons are passed through a specimen and focused by a series of magnetic lenses. Because electron wavelengths are much shorter than light, the magnification and resolution of electron microscopes are greatly increased. Special techniques must be used for viewing the specimens because humans are not capable of seeing electrons. Instead the electrons are directed to a fluorescent screen or photographic film for viewing.

Electrons have a tendency to scatter. To counteract this, viewing with an electron microscope must be done through a vacuum. Therefore, specimens cannot be alive as with the light microscope. To prepare specimens, the cells are killed, fixed (to prevent decomposition), embedded in a matrix (for strength), thinly sliced and stained.

Other types of microscopes, useful in studying cells, include the phase-contrast, interference contrast, dark field illumination, scanning electron, and atomic force microscopes.

Proper Handling of the Light Microscope

Microscopes are very expensive items and care should always be used when handling. The following directions must be followed:

- Transport: Refer to Figure 4.2: When carrying the microscope, always use both hands. One hand should be held under the base the other on the arm of the microscope and close to your torso. Never carry two microscopes at one time.

- Clutter: Keep your work station free of clutter when doing microscope work. Unnecessary books and personal items may lead to accidents.

- Electric Cord: Microscopes may be inadvertently pulled off of lab benches when students get tangled in a dangling electric cord. Don't let the cord dangle; wrap the excess cord around the gas or vacuum valves.

- Lens care: At the beginning of each laboratory, check the lenses to be sure they are clean. Only use lens paper to clean the lenses.

- Storage: Wrap the electric cord around the holder, put the dust cover over and microscope and place it onto the shelf with the oculars facing outwards.

Components

Refer to Figure 4.3 to identify the principal parts of the compound light microscope.

- Framework: The basic frame structure includes the arm and base. Other parts are attached to this framework.

- Stage: The horizontal platform that supports the microscope slide is called the stage. Note that the stage has a clamping device, the mechanical stage, to hold the slide on the stage. The slide than can be moved on the stage by using the stage control knobs. The hole is called the iris.

- Light Source: In the base is positioned the light source.

- Lens System: The microscopes contains: the oculars, objectives, and condenser.
    - The ocular or eyepiece, found at the top of the microscope, consists of two or more internal lenses and usually has a magnification of 10X. (The magnification is printed on the lens, you are looking only for the number that is followed by an X.) A microscope can have two oculars (binocular), or one (unocular).

    - The objectives, of which there are generally three or more, are found attached to a rotatable nosepiece. Thus, you can position the objectives over a slide. Most laboratory microscopes have objectives with magnifications of 4 X, 10X, 40 or 45X, and 100 X, designated as scanning power, low power, high power, and oil immersion, respectively.

    - The condenser is a lens that serves to concentrate light from the illumination source that is in turn focused through the object and magnified by the objective lens

- A diaphragm lever controls the amount of light exiting from the condenser under the stage.

- Focusing Knobs: The outer concentric knob controls the coarse adjustment and the inner knob controls the fine adjustment for bringing objects into focus.

Directions for Focusing Your Microscope:

1. Rotate the 10X objective over the light source and set the iris diaphragm control lever so that it is pointing directly away from you.
2. Turn the coarse focus knob towards you until it stops.
3. While looking through the eyepiece, slowly rotate the course focus until the specimen comes into sharp focus. If necessary, use the fine focus to bring the specimen into sharp focus.
4. Adjust the iris diaphragm until you obtain maximum contrast.
5. If necessary, rotate the objectives to put the 40X lens into place. You will only need to use the fine focus knob. (The microscope is parafocal—once focused will be focused at all magnifications.)
6. For a new slide, you should only need to use the fine focus knob (by starting at step 3).

Determining the Size of a Microscopic Object:

Permanently mounded into one of the oculars is a pointer and an ocular micrometer (Figure 4.1). The ocular micrometer is calibrated so that the space between two lines will be:

40X = 25 μm
100X = 10 μm
400X = 2.5 μm

Figure 4.1: Ocular Micrometer

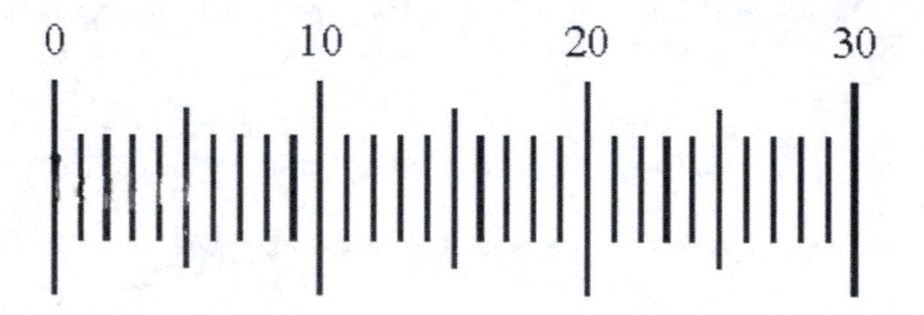

Figure 4.2: Holding a Microscope

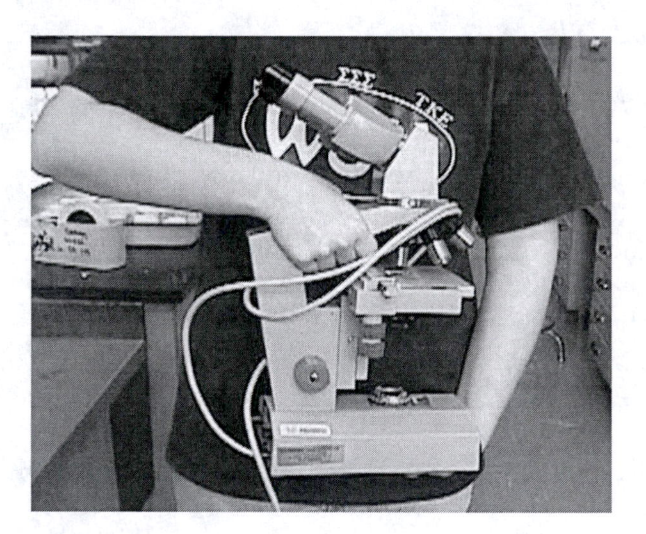

Figure 4.3: Parts of a Compound Light Microscope

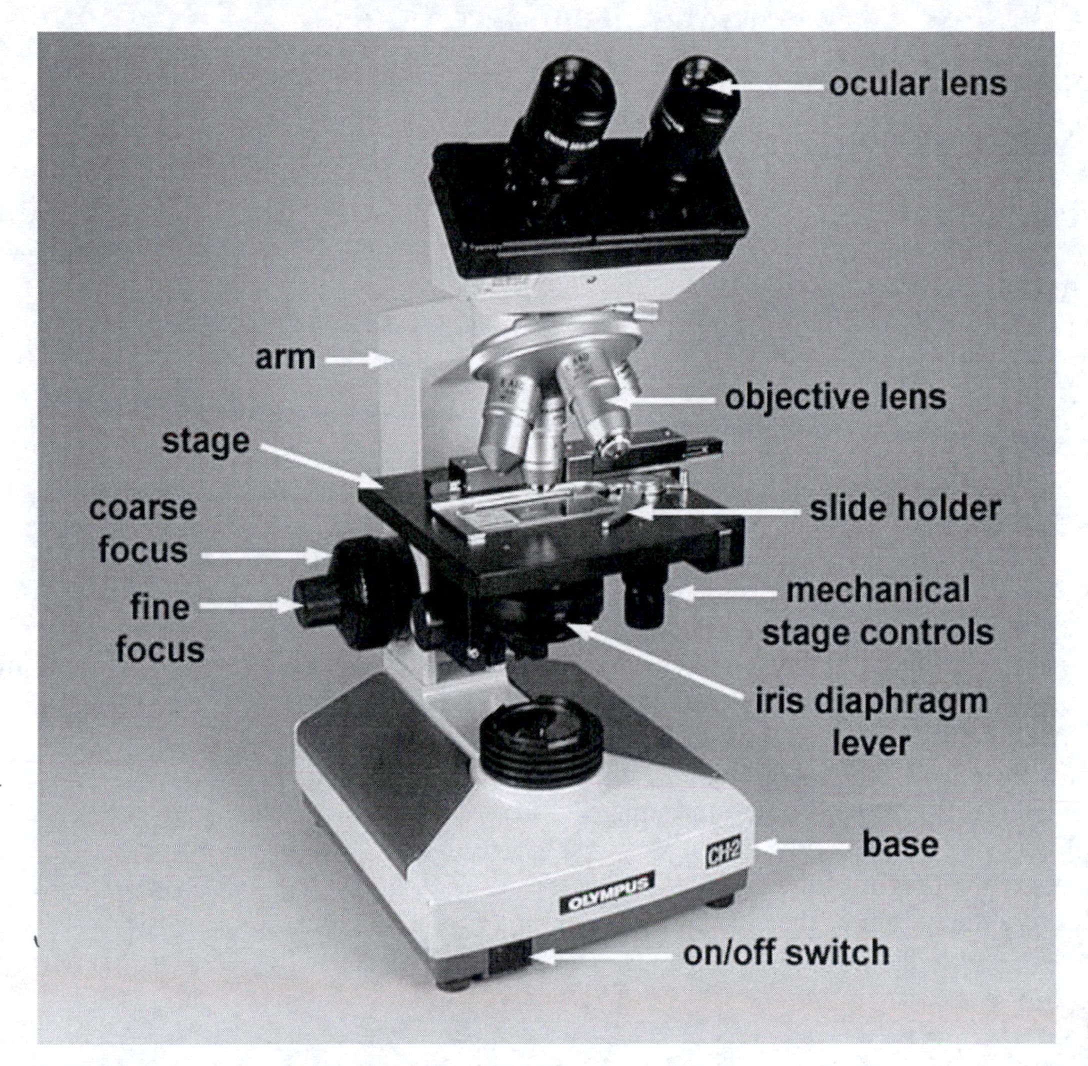

4x #spaces x 25 um

Materials and Methods

Focusing the Microscope

Materials

- compound light microscope
- "e" slide
- thread slide

Procedure

Low Power

1. Make sure that the low power lens has clicked in place.
2. Draw the e as it appears on the slide.
3. Place the "e" slide on the stage with the “e” centered. Place the clip on the slide to hold it in place. Use the stage control to move the slide, if necessary.
4. Focus the microscope following the directions on the previous page.
5. Draw the e as it appears through the eyepiece of the microscope.
6. Note the differences. Notice the image is inverted and reversed. Move the slide to the right.
7. Measure the height of the e.
8. Measure the height of the e using the scanning (40X) objective.

High Power

Compound light microscopes are parafocal, meaning that once the object is in focus with the lowest power, it should also be almost in focus with higher powers.

1. Have the object, the "e" in this case, into focus under lowest power as described above.
2. Move the slide until the "e" is centered in your field of view under the low power objective.
3. Move the high power objective into place by turning the objectives.
4. If any adjustment is needed, use only the FINE adjustment knob. Always only use the fine adjustment with high power. If you cannot find the e using the fine adjustment, return to the low power, refocus, and begin again.
5. On your drawing of the letter e, draw a circle around the portion of the letter that you are now seeing with the higher power. Notice that you see less of the image because your field of view has decreased.
6. Whenever you finish your observations of a slide, rotate the objective to the lowest power and then remove your slide.

Total Magnification

Total magnification is calculated by multiplying the magnification of the ocular lens by the magnification of the objective lens. Determine the total magnification of your microscope under low and high powers used above. Resolution (or resolving power) refers to the property of the microscope that shows two adjacent points on the magnified image as distinct entities.

Depth of Field

1. Place the slide containing the threads on the microscope.
2. Using the low power directions, move the objective as close as possible to the slide. Gradually bring the slide into focus. This will be the bottom thread.
3. Using the fine focus, determine which thread is in the middle and on the top.

Results

Focusing the Microscope

Low Power

Draw the "e" as it appears.

When the slide is moved to the right, which way does the image appear to move?

Determine the height of the e.

	Number of Spaces	Size
Scanning (40X) Objective	32	
Low Power (100X)	77	

High Power

Draw a circle around the portion of the e, sketched above, that is in your field of view at the higher power. When you look at the e using the higher power, you cannot see the entire letter. Why?

Total Magnification

	Ocular lens		Objective lens		Total
low power	10	X	4	=	40
high power	10	X	10	=	100

Depth of Field

Bottom thread: Yellow
Middle thread: Blue
Top thread: Red

Conclusions

1. Define the following: resolution, magnification, contrast, parafocal, ocular scale, depth of field and field of view.

2. Which of the following microscopes would be the best to view the following microscope choices: Brightfield, transmission electron, scanning electron
    a. Ribosomes
    b. Moving amoeba
    c. The indentations of a butterfly wing
    d. A skin sample stained with methylene blue
    e. Mitochondria

EXPERIMENT 5: CELL STRUCTURE

Objectives

1. Briefly describe the Cell Theory.
2. Identify on models or pictures of a eucaryotic the cell components that are listed in the methods section. State the function of each structure.
3. Explain the differences in organelles between plant and animal cells.
4. Prepare a stained slide of human cheek and onion skin cells. Using a light microscope, draw a cell and label the visible parts of each cell.

Introduction

Cell Theory

The cell theory states that all living organisms are composed of one or more cells; the cell is the smallest unit of living matter; cells are capable of self-reproduction; and cells come only from preexisting cells.

The cell theory grew out of thc thinking of many biologists. Beginning in the seventeenth century with the improvement of the first microscopes, Anton van Leeuwenhoek of Holland first observed tiny one-celled living things that no one had seen before. Robert Hooke, an Englishman, confirmed these findings, and was the first to use the term cell. Several nineteenth century biologists lent support to the cell theory, among them German biologists Scheleiden, Schwann, and Virchow, as well as Frenchmen Lamarck and Dutrochet.

Modern biologists recognize two types of cells: prokaryotic and eukaryotic. Prokaryotic cells lack a nuclear envelope and membranous cytoplasmic organelles. Single-celled bacteria are the only organisms that are prokaryotic cells. Eukaryotic cells have a true nucleus, a membrane-bound compartment that houses DNA within chromosomes. All cells, with the exception of bacteria, are eukaryotic. Eukaryotic organisms include the algae, protozoa, and fungi, as well as plants and animals.

Some cells are whole organisms, such as bacteria, protozoans, and most algae. In other words, they carry on all the functions of an organism: respiration, digestion, excretion, etc., and they are independent of other cells. On the other hand, many cells live cooperatively with others in a tissue and cannot survive in isolation, such as those in plants and animals. Here, they have sacrificed independence for efficiency. Multicellular organisms can adapt to diverse conditions in the environment where single celled organisms cannot.

Cell Structure and Function

All cells are bound by a plasma membrane, which regulates the passage of materials in and out of the cell.

Eukaryotic cells have a nucleus and various organelles (specialized membrane-bound structures). The nucleus is the prominent structure in the cell. The nucleus contains chromosomes, which are composed of DNA, and is surrounded by the nuclear envelope. The DNA contains the directions to make proteins needed by the cell. A specialized region of the nucleus, known as the nucleolus, appears as a darker staining region and is the site of ribosomal RNA production.

The ribosomes are the smallest structure and are involved in protein synthesis. They are comprised of ribosomal RNA and protein.

The group of organelles known as the Endomembrane System includes the endoplasmic reticulum, the Golgi complex, and various type of vesicles such as lysosomes and vacuoles. The Endoplasmic Reticulum (ER) is a system of membranous flattened channels and tubular canals. Some regions, known as the rough ER, are studded with ribosomes and function in protein synthesis. The ER serves also as a transport system for proteins and other substances.

The Golgi complex, a stack of membranes, functions in processing, packaging, and secretion of molecules. Lysosomes are membranous sacs produced by the Golgi complexes that contain specific hydrolytic enzymes. Vacuoles and vesicles are membranous sacs that serve as storage sites.

The organelles of energy are the mitochondrion and the chloroplast. Mitochondria are the site of cellular respiration where large quantities of ATP (useable energy for the cell) are made. Chloroplasts are found in plant cells and are the site of photosynthesis.

The cytoskeleton, comprised of microtubules and microfilaments, is responsible for the cell's shape and motility, and includes cilia, flagella, and the centriole. Cilia and flagella provide the movement of the cell. The centriole is involved in cell division.

Materials and Methods

Cell Structure and Function

Materials

- Animal Cell Model
- Plant Cell Model

Procedure

1. Identify and give the function of the following structures on the animal and plant models: (Use your textbook illustrations as a guide.) cell wall, plasma membrane, nucleus, nuclear envelope, nucleolus, mitochondria, chloroplast, smooth ER, rough ER, Golgi complex, ribosomes, vacuoles, centrioles, lysosomes.

Animal versus Plant Cells

Materials

- safety goggles
- gloves
- microscope slides
- cover slips
- toothpicks
- staining solution (methylene blue or IKI),
- onion

Procedure

Human Epidermal Cells

1. Gently scrape the inside of your cheek with a clean, flat toothpick.
2. Place the scrapings on a clean, dry slide.
3. Add a drop of stain.
4. Rinse under gently running water. Air dry.
5. Observe under the microscope at low and high.
6. Using high power, sketch and label one cell.
7. Determine the length and width of one cell.

Onion Epidermal Cells

1. With a sharp knife or scalpel, strip a small, thin transparent layer of cells from the inside of an onion leaf.
2. Place it gently on a clean glass slide.
3. Add a drop of stain.
4. Rinse under gently running water.
5. Cover with a cover slip, and observe under low.
6. Locate a cell with a nucleus.
7. Using low power, sketch and label one cell.
8. Determine the length and width of one cell.

Results

Cell Structure and Function

Identify each structure below on the animal and/or plant model and give its function.

STRUCTURE	**FUNCTION****
Cell Wall	______________________________
Plasma Membrane	______________________________
Nucleus	______________________________
Nuclear Envelope	______________________________
Nucleolus	______________________________
Mitochondria	______________________________
Chloroplast	______________________________
Rough Endoplasmic Reticulum	______________________________
Smooth Endoplasmic Reticulum	______________________________
Golgi Complex	______________________________
Ribosomes	______________________________
Vacuoles	______________________________
Centrioles	______________________________
Lysosomes	______________________________

**For each structure note if it is found in the animal cell, plant cell, or both.

Animal versus Plant Cells

Human Epidermal Cells

Sketch (fill the space) and label the nucleus, cell membrane, cytoplasm and bacteria of one cheek cell as it appears under high power. Indicate the cells length and width of one cell.

Onion Epidermal Cells

Sketch (fill the space) and label the cell wall, cell membrane, nucleus and cytoplasm of one onion cell as it appears under low power. Indicate the length and width of one cell.

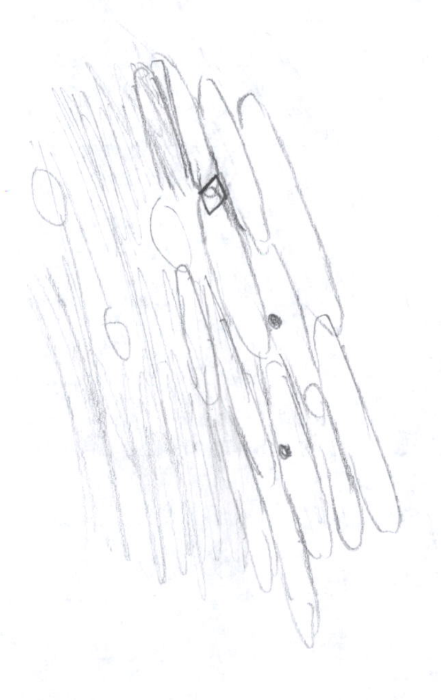

Describe some of the differences that you observed between onion (plant) and cheek (animal) cells.

Differences	**Plant Cell**	**Animal Cell**
Shape		
Cell Wall		
Other		

Conclusions

1. How does the cell wall differ in structure (molecular composition) and function from the plasma membrane?

2. Why are the onion cells all the same size and shape and the cheek cells all different?

3. The cheek cells and onion cells came from epithelial tissue. However, when observed under the microscope the onion cells were in a sheet and the cheek cells were individual units. What caused this difference?

4. Name two special organelles/structures that are associated with plant cells.

5. How does the nucleus control the cell?

EXPERIMENT 6:
DICHOTOMOUS KEY FOR PROTOZOA

Objectives

1. To use a dichotomous key to identify protozoa.
2. Explain how unicellular organisms can perform all the necessary functions of life and how each has unique adaptations.

Introduction

To classify organisms, you must first identify them. A taxonomic key is a device for identifying an object unknown to you but that someone else has described. The user makes choices between a set of alternative characteristics of the unknown object and by making the correct choices will arrive at the name of the object.

Keys that are based upon successive choices between two alternatives are known as dichotomous keys. When using a key, always read both choices, even though the first appears to be the logical one. Do not guess at measurements; use a scale. Since living organisms vary in their characteristics, do not base your conclusion on a single specimen if more are available.

Materials and Methods

Materials

- culture tube/slide containing a protist
- special slide holder
- microscope

Procedure:

Adapted from Carolina Biological

Take one of the numbered culture tubes that contains a protist and using the key and pictures identify the organism that is found in each tube. Use low power. Also show the "number" path that you followed to name the organism.

1. White or colorless 2
   Colored.. 8

2. Slow creeping (sliding), floats, no apparent motion 3
   Exhibits other motion.. 7

3. Spherically shaped with radiating spines .... ACTINOSPHAERIUNM
   Not spherical in shape.. 4

4. Shape remains constant.. 5
   Shape constantly changes 6

5. Flat test or shell with no embedded or attached material, pale to brown color.. ARCELLA
   Dome shaped test or shell with attached particles (usually sand).. DIFFLUGIA

6. Small, creeps using pseudopodia (false feet); single, disc shaped nucleus.. AMOEBA
   Large, creeps using pseudopodia; 100 small nuclei.. PELOMYXA

7. Cell has hair like structures (cilia) 16
   Cell's organ of locomotion is long whip-like flagella 9

8. Green color.. 9
   Color not green 25

9. Colony of many cells 11
   Single, motile cell 10

10. One observed flagella 15
    Two flagella 14

11. Flat, disc colony of 16 cells .... GONIUM
    Spherical colony.. 12

12. Colony of 32 or fewer cells..13
Colony or more than 32 cells...VOLVOX

13. Colony of 32 cells...EUDORINA
Colony or 16 cells...CHLAMYDOMONAS

14. Elongated cell with narrowed posterior...CHILOMONAS
Oval shaped cell...CHLAMYDOMONAS

15. Elongated, green cell...EUGLENA
Elongated, colorless, with broad, round or truncated posterior; highly plastic
....PERANEMA

16. Cell entirely covered with cilia...20
Cilia in specialized or localized areas...17

17. Cell or stalk and attached to debris...18
Cell not on stalk...19

18. Stalk not branched; stalk contracts like a spring...VORTICELLA
Stalk branched; stalk appears to contract like a spring..............CARCHESIUM or EPISTYLIS

19. Oval shaped cell with distinct point like projections used for "walking"
....EUPLOTES
Oval shaped cells with 2 ciliary bands, swims in spiral motion................................DIDINIUM

20. Body trumpet shaped or elongated...21
Body oval shaped...24

21. Body trumpet shaped; usually attached to substrate...STENTOR
Body elongated; never attached to substrate...22

22. Body elongated with distinguishable trunk...DILEPTUS
Body elongated without trunk...23

23.Large with elongated, flat body with blunt ends, contracts to ¼ its length
....SPIROSTOMIUM
Small, "cigar" shaped, swims in corkscrew fashion...PARAMECIUM

24. Small, oval with small mouth; fast swimmer...COLPIDIUM
Extremely large (visible with naked eye), with large, wide mouth...........................BURSARIA

25. Pink or rose; cilia...BLEPHARISMA
Dark, bluish-green; cilia...STENTOR

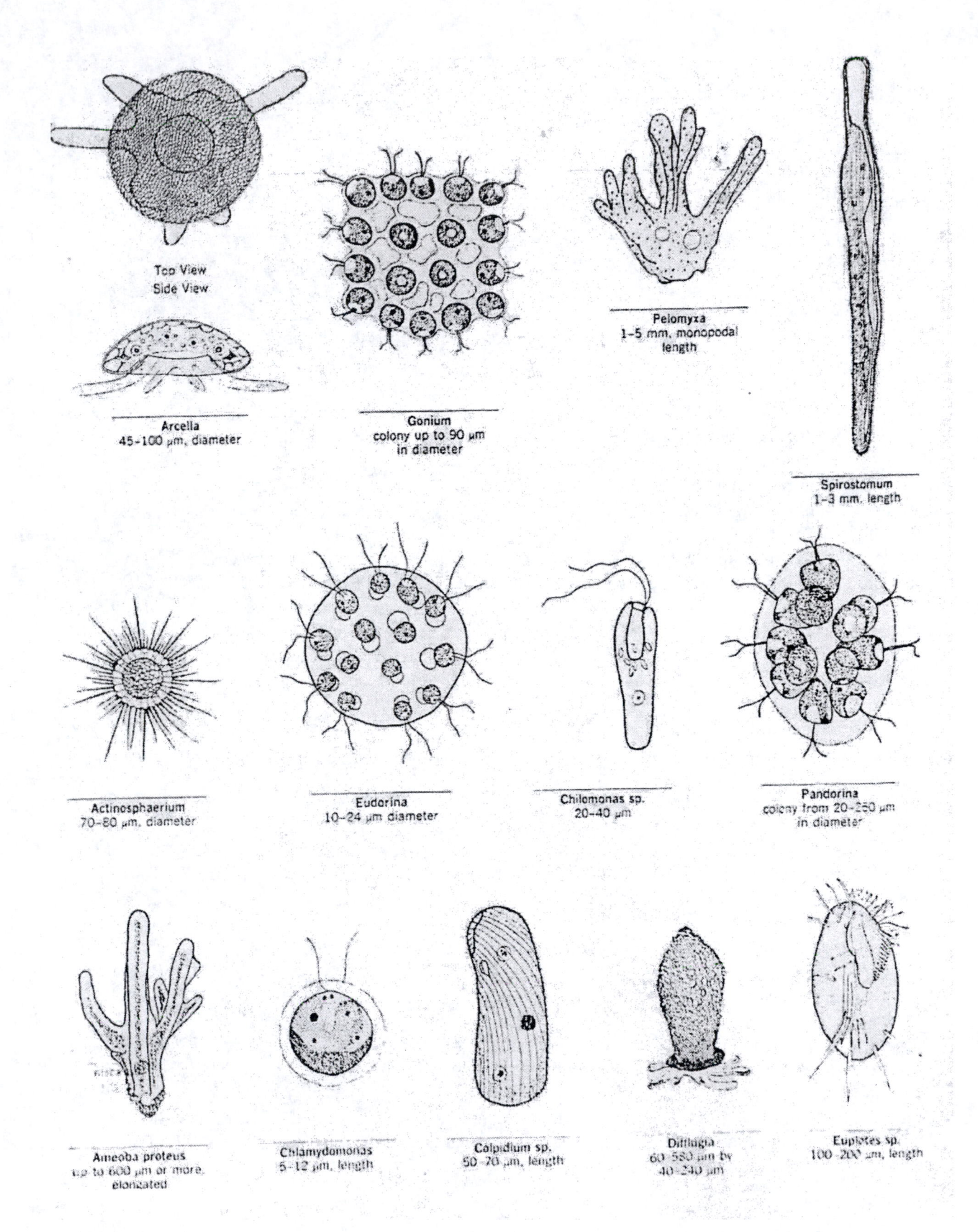
Top View
Side View
Arcella
45–100 μm, diameter
Gonium
colony up to 90 μm
in diameter
Pelomyxa
1–5 mm, monopodal
length
Spirostomum
1–3 mm, length
Actinosphaerium
70–80 μm, diameter
Eudorina
10–24 μm diameter
Chilomonas sp.
20–40 μm
Pandorina
colony from 20–250 μm
in diameter
Ameoba proteus
up to 600 μm or more,
elongated
Chlamydomonas
5–12 μm, length
Colpidium sp.
50–70 μm, length
Difflugia
60–580 μm by
40–240 μm
Euplotes sp.
100–200 μm, length

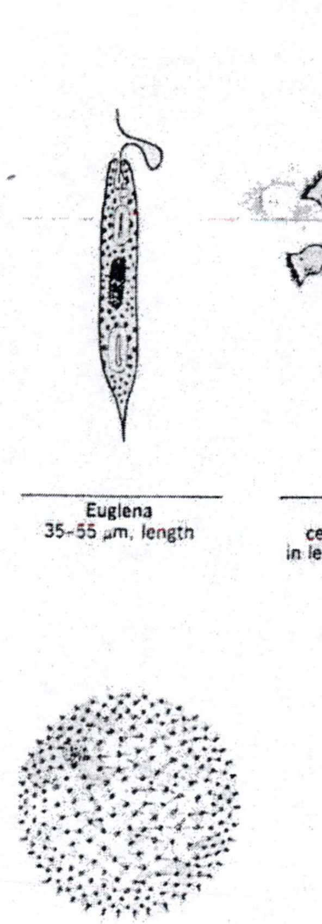

Euglena
35–55 μm, length

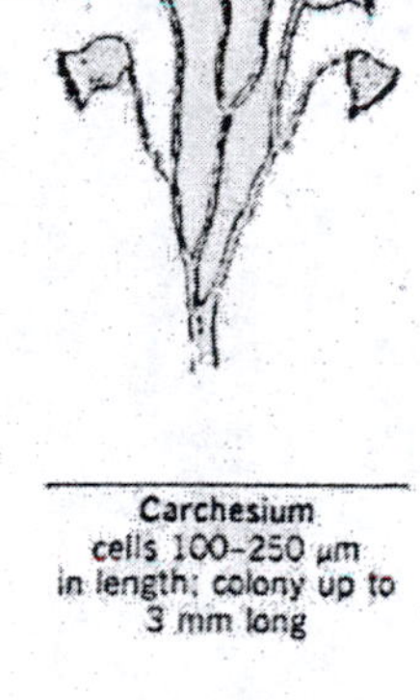

Carchesium
cells 100–250 μm in length; colony up to 3 mm long

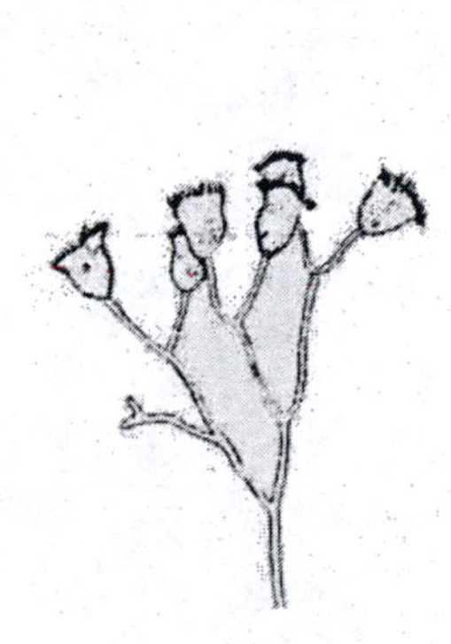

Epistylis
cells 50–100 μm in length; colony up to 3 mm long

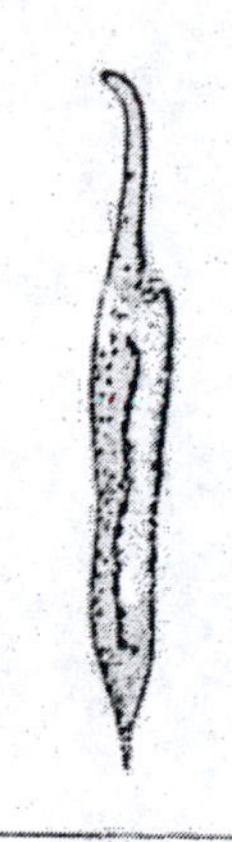

Dileptus
250–400 μm, length

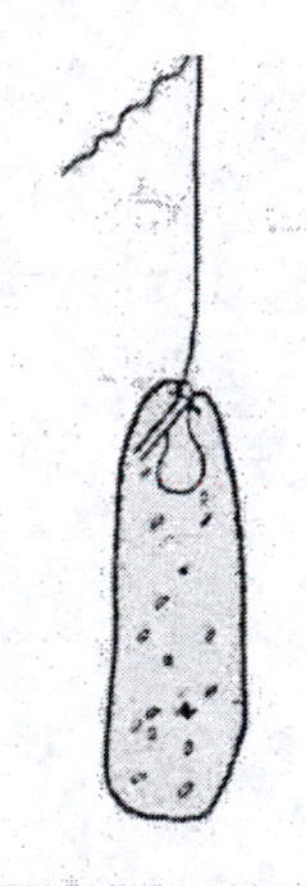

Peranema sp.
20–70 μm, length

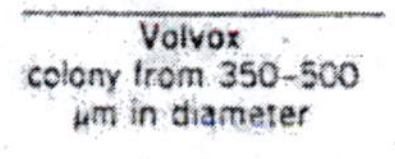

Volvox
colony from 350–500 μm in diameter

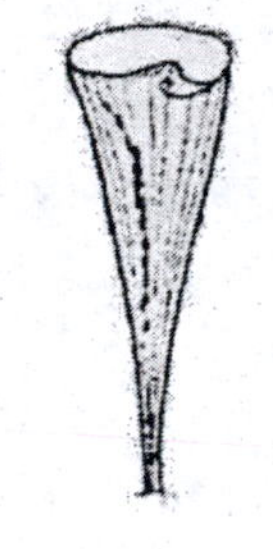

Stentor coeruleus
1–2 mm, extended

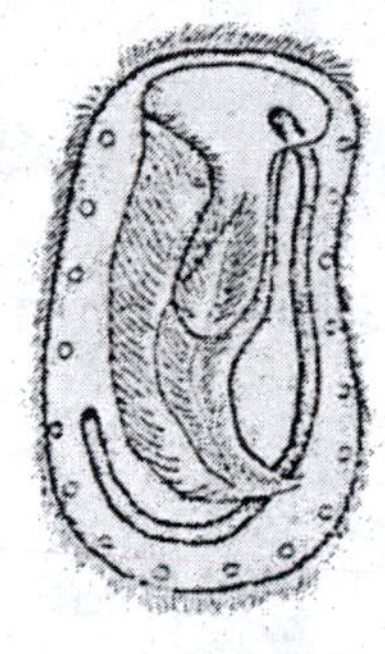

Bursaria truncatella
500–1000 μm, length

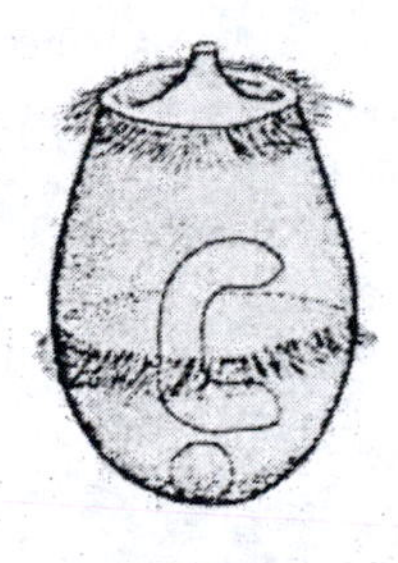

Didinium
80–200 μm, length

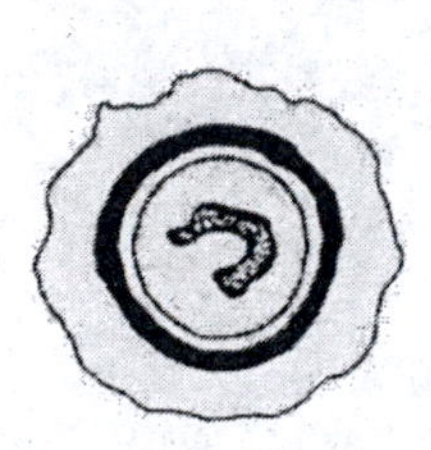

Didinium Cyst

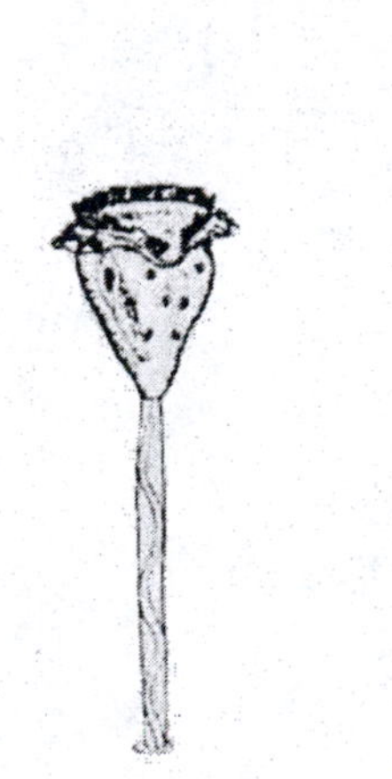

Vorticella
50–145 μm, length

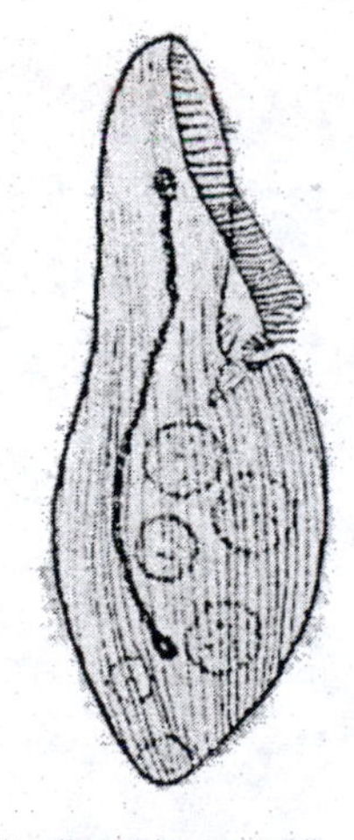

Blepharisma sp.
400–600 μm, length

Paramecium Species

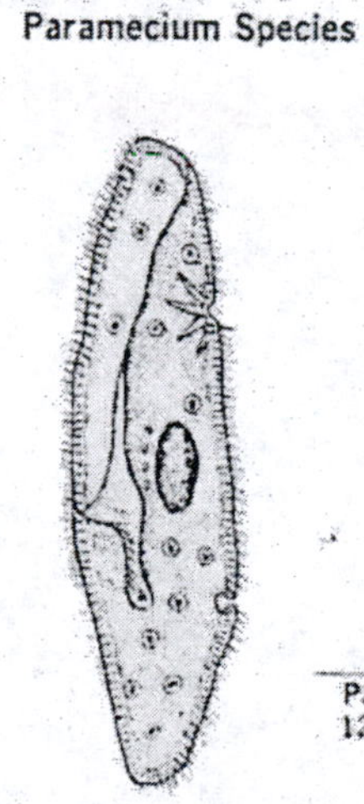

Paramecium multimicronucleatum

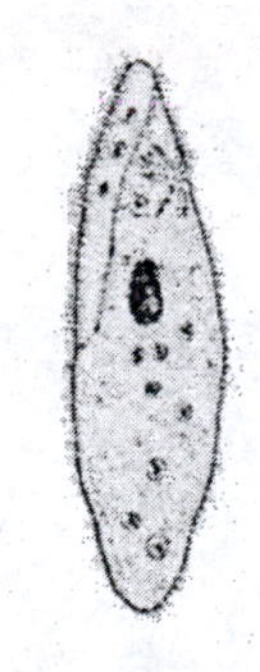

Paramecium aurelia
120–180 μm, length

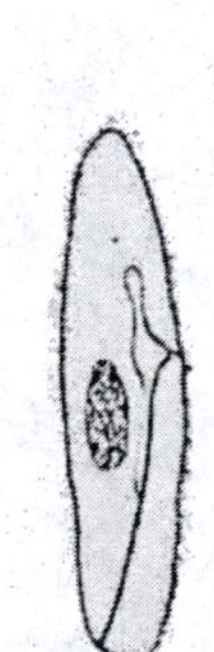

Paramecium caudatum
180–300 μm, length

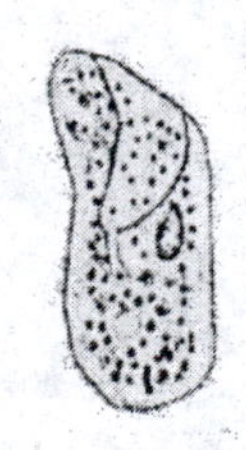

Paramecium bursaria
70–110 μm, length

Results

Organism #	Path Followed	Identity

Conclusions

1. Protozoans are described as single celled organisms that exhibit both plant-like and animal-like properties. Name a plant-like and an animal-like characteristic that you observed in the protists.

2. Construct a dichotomous key that can be used to differentiate the following shapes, A,B,C,D,E.

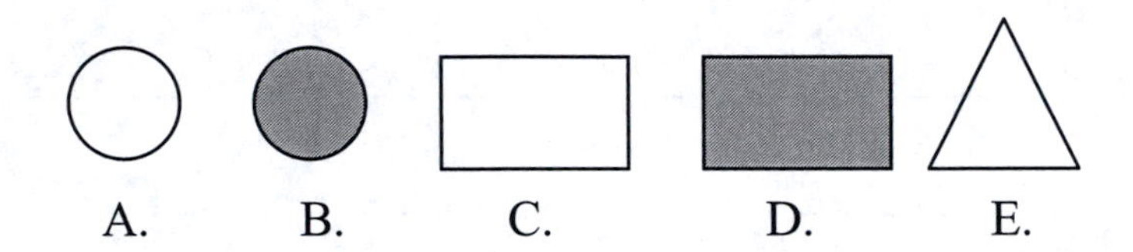

EXPERIMENT 7: DIFFUSION AND OSMOSIS

Objectives

1. Explain the difference between active and passive transport.
2. Define diffusion and osmosis.
3. Define solution.
4. Describe how osmosis occurs when an animal and plant cell is placed into hypotonic, hypertonic or isotonic solutions.
5. Using osmosis, determine the sugar content of an egg.
6. Explain how a concentration gradient affects osmotic pressure.
7. Define the term selectively permeable and state how it applies to the plasma membrane.

Introduction

Diffusion

Diffusion is the movement of molecules from an area of greater concentration to an area of lesser concentration. In other words, a substance will diffuse down its concentration gradient.

Some of the traffic across the membranes of cells occurs by diffusion. Whenever a substance is more concentrated on one side of a membrane than on the other, there is a tendency for the substance to diffuse across the membrane down its concentration gradient. The diffusion of a substance across a biological membrane is called passive transport. The cell does not have to expend energy for a substance to move across a membrane by passive transport; the molecules are simply diffusing down their concentration gradient, which is a spontaneous process caused by the natural movement of molecules, called Brownian Motion. The particles will naturally move to areas of lower concentration because the likelihood of collisions is greater in a higher concentration. The particles move in a straight line until they bump into something thus changing their direction of movement. Keep in mind that each particle is exhibiting random motion (cannot predict which way an individual particle will move) but you can predict the net movement of the collection of particles.

However, biological membranes are selectively permeable; that is, they allow some substances to freely cross more easily than others. Thus, the cell is able to retain many varieties of small molecules and exclude others. Moreover, substances that move through the membrane do so at different rates. Hydrophobic or non-polar molecules, such as hydrocarbons, oxygen, and carbon dioxide can dissolve in the membrane and cross it with ease due to the hydrophobic nature of the membrane. Conversely, most hydrophilic molecules, such as ions and polar molecules cannot pass through the membrane. However, very small polar molecules that are not charged can pass through. An example is water, which is small enough to pass between the lipids of the membrane.

Osmosis

Osmosis is a special case of passive transport that deals with the diffusion of water across a selectively permeable membrane. Osmosis depends upon the concentration of solutes in a solution. A solution is comprised of a solvent, the substance that does the dissolving, and a solute, the substance that is being dissolved. In the case of osmosis, the solvent is water. The solute is the substance, such as a sugar or salt that is dissolved in the water.

In comparing two solutions of unequal solute concentration, the solution with a greater concentration of solutes is said to be hypertonic. The solution with the lesser solute concentration is hypotonic. These are relative terms that are only meaningful in a comparative sense. For example, tap water is hypertonic to distilled water, but hypotonic to seawater. In other words, tap water has a higher concentration of solutes than distilled water, but a lower concentration than seawater. Solutions of equal solute concentration are said to be isotonic.

If a cell is placed in an environment which is isotonic, that is to say an environment of equal solute concentration, then there is no concentration gradient. Without such a gradient there will be no net flow of water across the membrane.

If a cell is placed in a hypotonic environment, then there is a lower solute concentration surrounding the cell than within the cell. Whenever the solute concentration is lower, the concentration of free water molecules is higher. Therefore, water will diffuse down its concentration gradient from an area of higher concentration (outside the cell) to an area of lower concentration (inside the cell). So, when placed in this hypotonic environment, a cell will swell, and possibly burst.

A cell, which is placed in a hypertonic environment, is surrounded by a solution, which is higher in solutes, hence lower in free water molecules. In this case, the free water concentration is greater inside the cell than outside, so the water flows out of the cell. In a hypertonic environment a cell will shrink and possibly dehydrate.

Remember: The water always moves <u>into</u> the hypertonic solution.

Osmotic Pressure is the pressure that must be exerted on the hypertonic side of the semi-permeable membrane to prevent osmosis. The greater the concentration difference across the membrane, the larger the gradient and hence the greater the osmotic pressure.

Materials and Methods

Diffusion of a Solid in a Liquid

Materials

- safety goggles
- potassium permanganate crystals
- 100 ml beaker half filled with water

Procedure

1. Drop a single crystal of potassium permanganate into a beaker of water.
2. Do not disturb the beaker for the duration of the lab period.
3. Observe the beaker at the beginning of the experiment and at intervals throughout the laboratory.
4. Record your observations.

Osmosis (Decalcified Egg Method)

Adapted from Cocanour, B. and A.S. Bruce. JCST: Nov. 1985.

Materials

- safety goggles
- 4 decalcified eggs
- 4 beakers
- balance
- weigh boat
- plastic spoon
- solutions
    - 2.0 M glucose
    - 1.5 M glucose
    - 0.5 M glucose
    - water

Procedure

1. Eggs have been soaking in vinegar (4.0% acetic acid) for at least 72 hours. This removes the calcium shell leaving behind a soft semi-permeable membrane.
2. Select 4 eggs and blot them on a paper towel to remove excess vinegar.
3. Determine and record the mass of each egg in grams.
4. Add the following solutions to a beaker (3/4 full): distilled water, 0.5M glucose, 1.5M glucose and 2.0M glucose
5. Submerge an egg in each beaker and take its mass every 15 minutes (after blotting off excess solution with a paper towel) for 90 minutes. Record the new mass.
6. Calculate the % change in mass for each egg:

$$\frac{[\text{(new mass - initial mass)}]}{[\text{ initial mass }]} \times 100\%$$

7. Graph time (x) vs. % change (y) on a graph. All 4 eggs are on the same graph.

Osmosis in Plants

Materials

- elodea leaf
- microscope
- microscope slide
- cover slip
- concentrated NaCl solution with dropper

Procedure

Elodea is a common water plant. When a leaf of the plant is observed by means of a microscope, cells can be readily seen. Within the cells, numerous green chloroplasts are equally distributed. Under normal conditions, the cell membrane cannot be seen since it is pushed up against the cell wall. We will be looking for changes in this normal appearance as a result of osmotic events.

1. Place an Elodea leaf (young and green) in a drop of water on a microscope slide and place a cover slip over the leaf. Be careful to exclude air bubbles.
2. Observe the leaf under low. Draw and label one cell. Determine the width and length of one cell.
3. Remove the cover slip. Place one drop of concentrated salt solution on the leaf. Replace the cover slip.
4. Draw and label one cell. (Pay special attention to the change in distribution of the chloroplasts.

Results

Diffusion of a Solid in a Liquid

Sketch and label the appearance of the beaker at the start, middle and end of the lab period.

Osmosis - decalcified eggs

Table 1: Mass of Eggs

Egg in	**0 min.**	**15 min.**	**30 min.**	**45 min.**	**60 min.**	**75 min.**	**90 min.**
Water							
0.5 M glucose							
1.5 M glucose							
2.0 M glucose							

Table 2: % change in Eggs

Egg in	0 min.	15 min.	30 min.	45 min.	60 min.	75 min.	90 min.
Water							
0.5 M glucose							
1.5 M glucose							
2.0 M glucose							

Osmosis (Elodea Leaf)

Sketch (fill the space) the leaf before and after adding the salt solution. Label the cell wall, cell membrane and chloroplasts. Indicate the length and width of the cell.

Conclusions

Diffusion of a solid in a liquid

1. Explain how diffusion went from an area of high concentration to low concentration.

Decalcified egg osmosis

1. Using the graph of your data, describe each solution as to its tonicity. (Hypotonic, hypertonic or isotonic to the egg). Be sure to explain how you arrived at your conclusions—support with data.

2. Presuming the contents of the eggs were identical, what is the sugar content of the eggs? Explain (using your data) how you came to your conclusion.

3. Which solution created the most osmotic pressure on the decalcified egg?

4. Describe how the decalcified egg acts as a semi-permeable membrane.

5. To prepare 1000 ml of the following, how much glucose is needed: (Refer to the Sackheim chapter on Liquid Mixtures.) Show your work.
    a. 2 M glucose
    b. 1.5 M glucose
    c. 0.5 M glucose

Osmosis in Elodea leaf

1. Why did the overall shape of the plant cell remain the same after the leaf was placed into a concentrated salt solution?

2. What would happen to the elodea cell if it was placed into distilled water?

Compare and Contrast Diffusion and Osmosis

	Diffusion	**Osmosis**
Is a Gradient Needed?		
Direction of Movement (In relation to the solute)		
Is a Membrane Required?		
Source of energy for movement		
What is moved		

General Osmosis Review

For each picture, draw an arrow to indicate the flow of water. On the line in each beaker, indicate whether the solution in the beaker is isotonic, hypertonic or hypotonic.

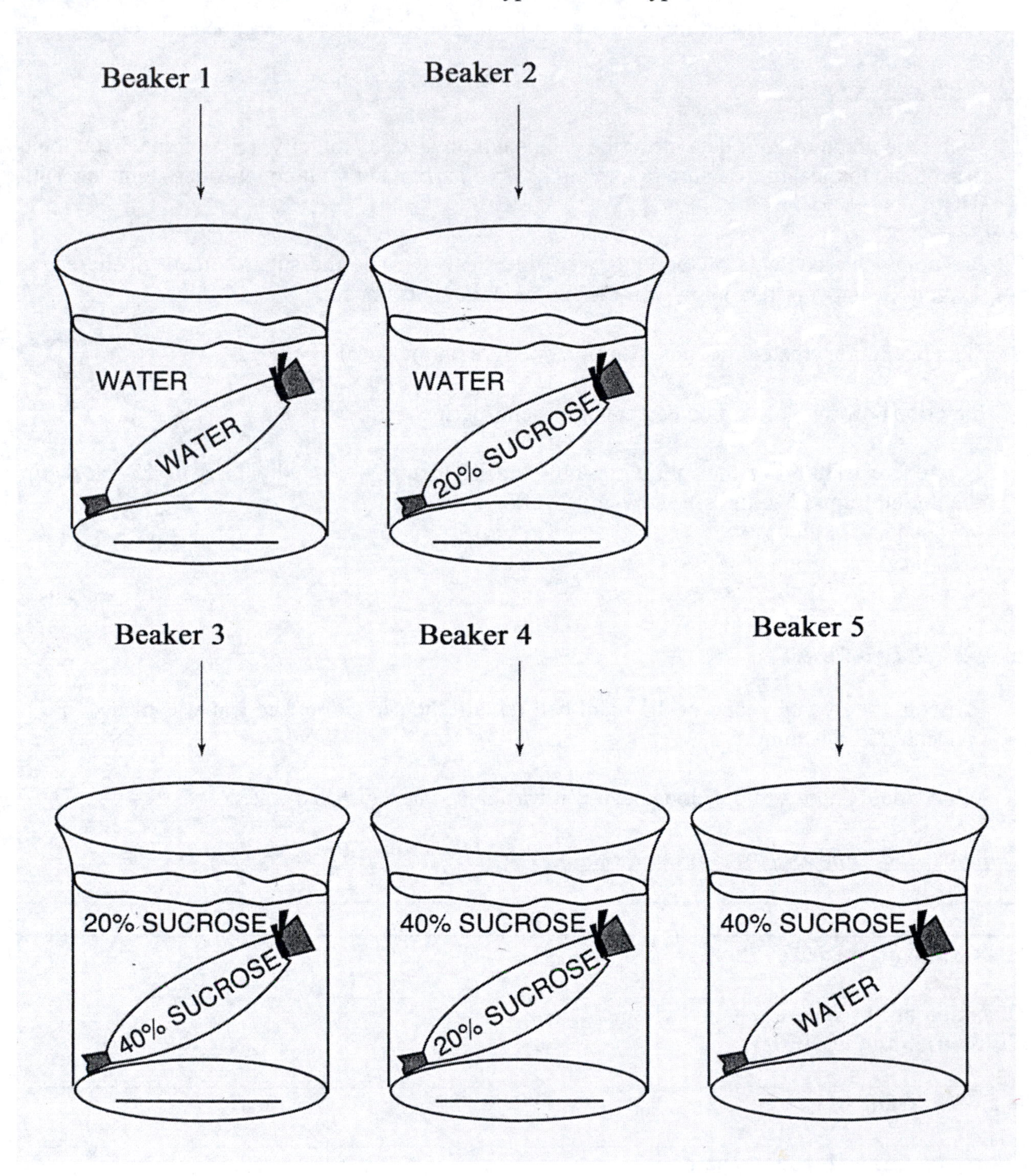

EXPERIMENT 8:
ENZYMES

Objectives

1. State the generalized reaction for an enzyme and substrate.
2. State the complete chemical reaction for the enzyme catalase.
3. Using absorbency data, calculate a reaction rate.
4. Determine the effects of enzyme availability on reaction rate.
5. Explain the effect of extremes of pH and temperature on reaction rates.
6. Explain the effect an increase in temperature has on most enzymes.
7. Describe what is meant by denaturing and relate this to the primary, secondary and tertiary structure of a protein.

Introduction

Enzyme Function

Every living organism carries out many chemical reactions. It is vital to the organism that these reactions occur at an extremely rapid rate, but a safe temperature. All organisms use enzymes to speed up the rate of chemical reactions without increasing the temperature. Enzymes are nearly always proteins with specific tertiary structures. They are also susceptible to denaturing (a disruption of the chemical bonds which maintain their integrity), as any other protein, when exposed to extremes in the environment.

Enzymes do not start a reaction; they merely speed up the reaction that is already under way. Without this increased rate of reaction, life as it occurs on this earth would not be possible. Each reaction in a cell requires a specific enzyme to allow the reaction to proceed at the proper rate. There are hundreds of different reactions necessary in the life of the cell, therefore hundreds of different enzymes are present in the cell. For an enzyme to work in a reaction, it must fit with a substrate or substrates. Each type of enzyme has a specific physical shape that determines its specificity.

When an enzyme's active site reacts with a substrate, the two molecules physically combine. The substrate is changed into the new product, but the enzyme is not changed by the reaction.

A general equation for an enzymatic reaction is:

$$\text{substrate(s)} + \text{enzyme} \rightarrow \text{end product (s)} + \text{enzyme}$$

or

$$\text{substrate} \xrightarrow{\text{enzyme}} \text{reactant(s)}$$

(Since the enzyme is not required nor used up in the reaction, it is often placed on top of the arrow.)

Catalase Reaction

The normal function of catalase is to convert toxic H_2O_2 (hydrogen peroxide), which can be produced in certain metabolic reactions, into harmless H_2O (water) and O_2 (oxygen).

The specific reaction for today's exercise is:

$$H_2O_2 + \text{catalase} \rightarrow H_2O + O_2 + \text{catalase}$$

or

$$H_2O_2 \xrightarrow{\text{catalase}} H_2O + O_2$$

This reaction can be measure by monitoring the production of oxygen. The amount of oxygen present after the reaction can be measured by measuring the accumulation of oxygen in a closed system or the appearance of chemically active oxygen.

Many dyes will react with active oxygen and the colorless dye will change to a colored state. This test is called a dye-coupled reaction. The enzyme-catalyzed reaction produces a product (oxygen) that enters into a secondary reaction with the dye. (The enzyme does not bind with nor affect the dye.) In your experiment, you will use the dye, guaiacol, which turns brown when oxidized. The entire reactions are as follows:

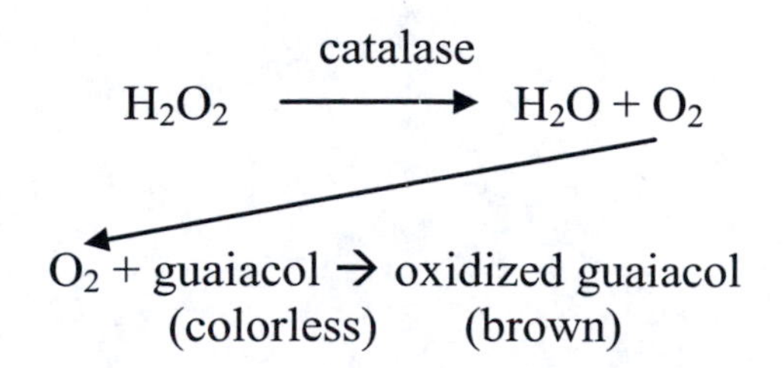

To quantitatively measure the amount of brown color in the final product, the enzyme, substrate, and dye can be mixed in a tube and immediately placed into a spectrophotometer that has been set to an absorbance of 500 nm.

In this exercise you will determine the effects of several factors on the activity of the enzyme, catalase.

Materials and Methods

Microscope Work

The reaction between catalase (which is found is most living cells) and hydrogen peroxide is often observed when hydrogen peroxide is added to bleeding cuts to disinfect the wound. The catalase in the blood cells reacts very quickly with the hydrogen peroxide and is observed as bubbles. In this exercise you will observe intact human blood cells. Human blood contains three type of cells: red blood cells (carry oxygen), white blood cells (fight infections) and platelets (clot the blood).

Obtain a prepared, stained human blood slide. Observe the specimen under high power. Find an area (under high power) that contains all three cell types. Draw and label one of each type of blood cell.

Catalase Experiment

Materials

- safety goggles
- gloves
- Enzyme preparation will be produced by instructor as follows: 5.0 g of turnip are ground and then blended in 500.0 ml of water. The mixture is then filtered.
- 11 cuvettes
- test tube rack(s)
- pipettes
- distilled water
- buffers: pH 3, 5, 7, 9
- hydrogen peroxide (H_2O_2) (Instructor: 3.33 ml of 3% H_2O_2 in 100.0 ml water)
- guaiacol
- thermometers
- ice bath: 0-4°C
- heating blocks: 37 °C, 48°C and 100 °C
- spectrophotometer (Review Experiment 3 on using a spectrophotometer.)

Experiment 1: Effect of Enzyme Availability

Procedure

1. Obtain 7 cuvettes. Working from left to right, prepare the tubes as follows: (all values are ml's)

Tube	Distilled Water	H_2O_2	Enzyme	Guaiacol
1 (blank)	5.0	2.0	0.0	1.0
2	0.0	2.0	0.0	1.0
3	4.5	0.0	0.5	0.0
4	0.0	2.0	0.0	1.0
5	4.0	0.0	1.0	0.0
6	0.0	2.0	0.0	1.0
7	3.0	0.0	2.0	0.0

2. Using Tube 1, adjust the spectrophotometer to zero absorbance at 500 nm (The filter lever is to the left.)
2. This step must be continuous: mix and insert! Clean the outside of the tubes. Mix the contents of tubes 2 and 3—using a gloved hand, cover the top of the tube and invert three times to mix the solution.
3. Mixing should be completed within 10 seconds. Place the cuvette into the spectrophotometer. Read the absorbance at 20-second intervals from the start of the mixing and continue for 2 minutes. Record your results. **If the initial reading was higher than 0.3, have the instructor review your results.** You may need to redo it. Note the color of the solution. **(Negative absorbance numbers are recorded as O.)**
4. Repeat step 3 but use tubes 4 and 5.
5. Repeat step 3 but use tubes 6 and 7.
6. Clean the tubes and dry the exteriors with a Kim Wipe.

Experiment 2: Temperature Effects

Purpose

1. Obtain 11 empty cuvettes. Working from left to right, prepare the tubes as follows: (all values are ml's).

Tube	Distilled Water	H_2O_2	Enzyme	Guaiacol
1 (blank)	6.0	0.0	1.0	1.0
2 (0-4°C)	0.0	2.0	0.0	1.0
3 (0-4°C)	4.0	0.0	1.0	0.0
4 (room temp)	0.0	2.0	0.0	1.0
5 (room temp)	4.0	0.0	1.0	0.0
6 (37°C)	0.0	2.0	0.0	1.0
7 (37°C)	4.0	0.0	1.0	0.0
8 (48°C)	0.0	2.0	0.0	1.0
9 (48°C)	4.0	0.0	1.0	0.0
10 (100°C)	0.0	2.0	0.0	1.0
11 (100°C)	4.0	0.0	1.0	0.0

2. Pre-incubate all the solutions at the appropriate temperature for at least 15 minutes.
3. Using Tube 1, adjust the spectrophotometer to zero absorbance at 500 nm.
4. As you did in the 1st experiment, mix tubes 2 and 3 and take absorbance readings. Repeat for tubes 4 and 5, 6 and 7, 8 and 9, 10 and 11. Record the exact temperatures on the result sheet.
5. Clean the test tubes and use the Kim Wipes to dry the exterior.

Experiment 3: pH Effects

Purpose

1. Obtain 9 empty cuvettes. Working from left to right, prepare the tubes as follows: (all values are ml's).

Tube	Buffer—see tube number	H_2O_2	Enzyme	Guaiacol
1 (blank-water)	6.0	0.0	1.0	1.0
2 (pH 3)	0.0	2.0	0.0	1.0
3 (pH 3)	4.0	0.0	1.0	0.0
4 (pH 5)	0.0	2.0	0.0	1.0
5 (pH 5)	4.0	0.0	1.0	0.0
6 (pH 7)	0.0	2.0	0.0	1.0
7 (pH 7)	4.0	0.0	1.0	0.0
8 (pH 9)	0.0	2.0	0.0	1.0
9 (pH 9)	4.0	0.0	1.0	0.0

2. Using Tube 1, adjust the spectrophotometer to zero absorbance at 500 nm.
3. As you did in the 1st experiment, mix tubes 2 and 3 and take absorbance readings as you did in the 1st experiment.
4. Repeat for tubes 4 and 5, 6 and 7, 8 and 9.
5. Clean all the tubes and dry them by inverting them in the test tube rack.

Results

Microscope Work

Draw (fill the space) and label each of the three types of blood cells.

Catalase Experiment

Experiment 1: Effect of Enzyme Availability

Time Seconds	Tubes 2 & 3 0.5 ml enzyme	Tubes 4 & 5 1.0 ml enzyme	Tubes 6 & 7 2.0 ml enzyme
20	.010	-.004	.009
40	.011	.001	.017
60	.012	.001	.025
80	.010	.003	.032
100	.010	.007	.041
120	.010	0108	.048
Color of Solution	colorless	colorless	brown

Construct a best-fit straight-line graph of your data (X axis: time in seconds, Y axis: absorbance with the wavelength used). Plot all tests on the same graph. (Start with 20 seconds.)

Determine the activity value of each enzyme concentration by using the following calculation to find the slope of the line:

Enzyme Activity = Δ absorbance/Δ minutes

$$= \frac{\text{(absorbance at 90 seconds – absorbance at 30 seconds)}}{\text{(90 seconds – 30 seconds)}}$$

Determine your absorbance units from your graph not the data table!
In the space below, show your calculations and write the results in the chart.

Enzyme Amount	Enzyme Activity
0.5 ml	
1.0 ml	
2.0 ml	

Experiment 2: Temperature Effects

Time Seconds	2 & 3 -4 °C	4 & 5 37 °C	6 & 7 ____ °C	8 & 9 47 °C	10 & 11 100 °C
20	.036	.055		.061	-.003
40	.038	.056		.061	-.003
60	.039	.055		.061	-.004
80	.040	.056		.061	-.003
100	.040	.056		.061	-.004
120	.041	.056		.060	-.003

Construct a best-fit straight-line graph of your data (X axis: time in seconds, Y axis: absorbance with the wavelength used). Plot all tests on the same graph. (The points may plateau at the end.)

Determine the activity value of each temperature. In the space below, show your calculations and write the results in the chart.

Temperature	**Enzyme Activity**

Experiment 3: pH Effects

Time Seconds	Tubes 2 & 3 pH 3	Tubes 4 & 5 pH 5	Tubes 6 & 7 pH 7	Tubes 8 & 9 pH 9
20	-.006	-.02	-.009	.006
40	-.006	-.02	-.008	.009
60	-.006	-.02	-.008	.00
80	-.006	-.02	-.007	.00
100	-.006	-.02	-.006	
120	-.006	-.02	-.006	

Construct a best fit straight line graph of your data (X axis: time in seconds, Y axis: absorbance with the wavelength used). Plot all tests on the same graph. (The points may plateau at the end.)

Determine the activity value of each pH. In the space below, show your calculations and write the results in the chart.

pH	Enzyme Activity

Conclusions

1. What is the reaction (write the equation) that you are studying? Be sure to include the name of the enzyme. What was the origin of the enzyme?

2. What is causing the absorbency number to change? What is the purpose of the guiacol?

3. How did you measure the rate of the reaction? How did you know if the reaction rate changed?

4. For each experiment, you initially prepared two tubes but only one was placed into the spectrophotometer. Why?

5. Why did the reaction rate change when the amount of enzyme changed?

6. What was the optimal temperature for this enzyme? How did you arrive at your answer? Support with data.

7. What effect did temperature have on the reaction rate? Be sure to explain what caused the reaction rates to be similar at O°C and 100°C.

8. What was the optimal pH for this enzyme? How did you arrive at your answer? Support with data.

9. What effect did pH have on the reaction rate? How did you arrive at your answer? Support with data?

10. What effect would the age of the "vegetable" have on the experiment?

11. If you used a "cooked vegetable" to prepare the enzyme, what results would you expect? Why?

EXPERIMENT 9: PHOTOSYNTHESIS

Objectives

1. Explain the difference between autotrophs and heterotrophs.
2. Write the generalized equation for photosynthesis and state what cellular organelle is involved.
3. Explain how photosynthesis is the key process which allows heterotrophs to obtain energy from ingested plants.
4. Describe why the rate at which the leaf segments rise in this experiment can be used as a measure of the rate of photosynthesis. (Relate this to the overall reaction for photosynthesis, the solubility of oxygen and carbon dioxide in water, and the structure of a plant leaf.)
5. Describe how an environmental change can affect the rate of photosynthesis.

Introduction

Food and Energy

Food is required by all living organisms. It provides both building block molecules for synthesizing cellular materials and energy to drive cellular chemical reactions. The living world, with few exceptions, operates at the expense of energy captured by photosynthetic living organisms. They are able to build up the numerous complex molecules that contribute to their cellular structure and in other ways that are essential to their existence.

Plants and algae are autotrophs; they make their own food from carbon dioxide, water, and sunlight. Heterotrophs, like humans, must take in food that is already made. Hence, the ultimate source of energy for living organisms is the energy of sunlight that is captured and converted into organic molecules during photosynthesis. The overall reaction taking place in photosynthesis is given as:

$$\text{sun} + \underset{\text{carbon dioxide}}{6CO_2} + \underset{\text{water}}{6H_2O} \rightarrow \underset{\text{sugar}}{C_6H_{12}O_6} + \underset{\text{oxygen}}{6O_2}$$

Though this equation suggests otherwise, photosynthesis is a complex process involving many steps. These reactions can be divided into two groups: 1) The light (dependent) reaction, and 2) the light independent reactions (also called the Calvin Cycle). In the light reactions, light energy is absorbed by chlorophyll pigments and is converted into chemical energy (ATP). The light independent reaction, so named because it does not directly require light, utilizes the products of the light reaction to produce the end product of photosynthesis, a carbohydrate.

C3 Plant Leaf Structure

Plant cells which are active in photosynthesis contain green spherical cytoplasmic organelles known as chloroplasts. These organelles contain the pigments necessary for light absorption and the enzymes required for both the light dependent and light independent of photosynthesis. Photosynthesis occurs mostly in the palisade mesophyll cells which are located directly under the upper cuticle layer. (See Figure 9.1)

In large multicellular plants effective photosynthesis is possible as a result of effective transport mechanisms. Water is transported throughout the plant by the xylem, a complex tissue consisting of thick-walled and hollow cells resembling small pipes. Water is pulled through these cells and up the plant by the negative pressure generated as water evaporates from aerial portions of the plant. Water vapor leaves the plant through the stomata, which are small pores on the leaf surface. Gas exchange also occurs at the stoma (release of oxygen and uptake of carbon dioxide) and the gases are stored in the air spaces that are created by the spongy mesophyll layer of cells. The sugars are transported via the phloem. Phloem cells are thin-walled, yet shaped much like the xylem cells.

Figure 9.1: Leaf Cross Section

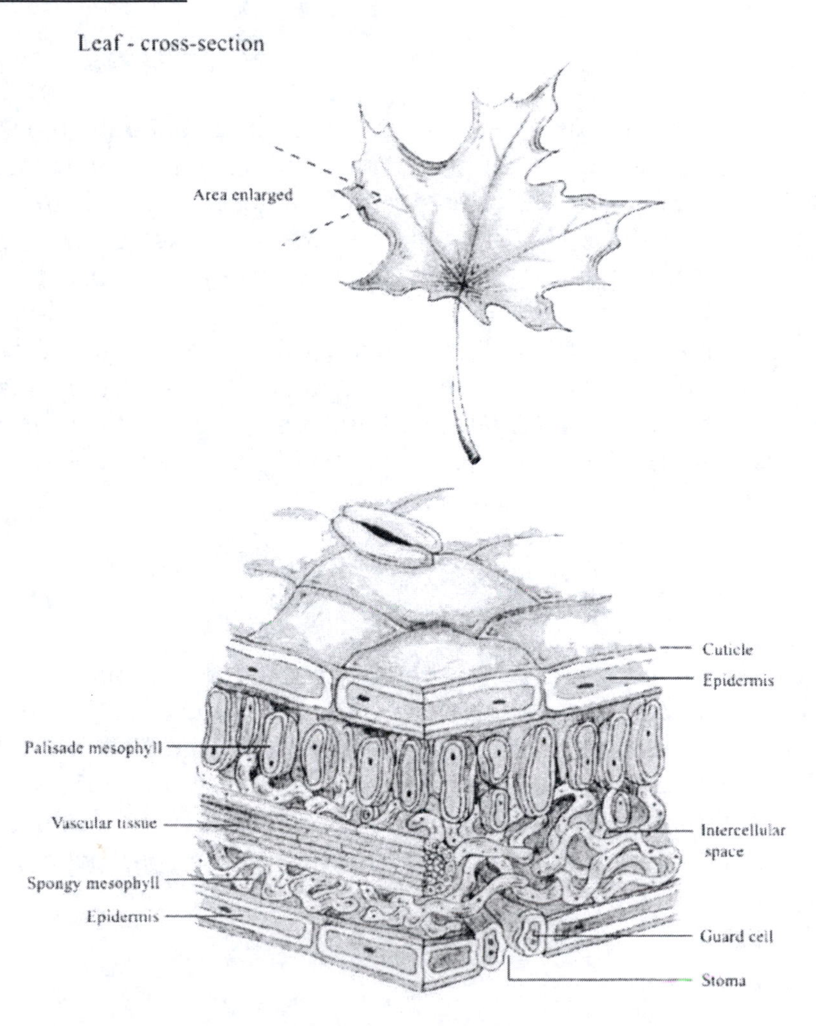

Photosynthesis Assays

There is no practical method to determine precisely the amount of sugar or oxygen produced during photosynthesis. However, several indirect methods can be used for this determination.

Carbon dioxide is highly reactive in water (To maintain the pH you will use the buffer, sodium bicarbonate. This solution will also be the carbon source because the gaseous CO_2 will react with the water and form carbonic acid. The buffer will convert the acid into sodium bicarbonate.) and plant cells readily absorb carbon dioxide from water solution during photosynthesis. On the other hand, oxygen, one of the end products of photosynthesis, is only sparingly soluble in water and most remains in the gaseous phase, accumulating in the plant's air spaces.

To measure the rate of photosynthesis in segments of spinach leaf, a vacuum pump will be used to remove the insoluble oxygen from the air spaces in the leaf and replace it with a bicarbonate infiltration solution. This will cause the segments to sink. (They would normally float due to the oxygen in the air spaces.) During photosynthesis, carbon dioxide in the water now filling the air spaces will be absorbed by the plant cells. In the presence of light, the carbon dioxide and water will be used to produce carbohydrates and oxygen will be given off (see the overall reaction for photosynthesis). Since the oxygen is only sparingly soluble in water, most will remain as gaseous oxygen. As the oxygen accumulates in the air spaces, the leaf segments will rise to the surface of the test bicarbonate solution. The more rapid the rate of photosynthesis, the faster the leaf segments will rise.

Materials and Methods

Microscope Work

The parts of a C3 leaf, described in the lab manual and given by your instructor, are easily observable using a microscope. Draw a cross section and label the following parts: spongy mesophyll cells, air spaces, palisade mesophyll cells, cuticle layer and stomata.

Spinach Leaf Assay

Adapted from a Carolina Biological

Materials

- safety goggles
- 600 ml beaker
- ice
- tray or beaker to collect leaf segments
- ring stand
- infiltration solution
- large culture tube
- bicarbonate solution
- goose neck lamp
- spinach leaves
- thermometer
- hole punch
- vacuum flask

Procedure

1. Prepare your experimental setup:
   Partially fill a 600ml beaker with water to serve as a heat sink (to absorb heat from the lamp) and set it on the base of a ring stand. Place a thermometer in the beaker of water.
2. Add bicarbonate solution to fill the culture tube about 3/4 full.
3. Set up a gooseneck lamp near the ring stand. Record the wattage of the bulb.
4. Using a round-hole punch or a cork borer, punch out 30 segments of spinach leaf and place them in a vacuum flask containing infiltration solution. All groups will place their leaf segments in the same vacuum flask.
5. Connect the flask to a vacuum pump. Evacuate for one minute and then disconnect and swirl for half a minute to break air bubbles. Repeat this vacuuming procedure two more times. (Do not over vacuum or you will rupture the plant cells.)
6. Pour the contents of the flask into a white enamel tray or beaker. Work in a dim light from here on to avoid a positive assay before the start of the experiment.

7. Place the lamp at a distance as instructed from the culture tube (not the beaker). Accurately measure and record this starting distance.
8. Select 10 segments of spinach leaf that have sunk to the bottom of the tray/beaker and add them to your culture tube containing the bicarbonate solution. Clamp the culture tube to the ring stand so the tube is three-fourths immersed in the beaker of water. Note the starting temperature.
9. Turn on the lamp and start timing. Note the time in minutes and decimal fractions of a minute (divide number of seconds by 60) it takes for each segment to rise to the top of the bicarbonate solution. During the experiment, keep checking the temperature of the water in the heat sink. As the temperature starts to rise, keep it constant by gradually adding a few ice chips. You don't want increasing temperature to influence the rate of photosynthesis.
10. Calculate the average time it took for the segments to rise.
11. Introduce a variable that will change the rate of photosynthesis. Each group will use a different variable. Repeat the experiment a second time using 10 fresh leaf segment and new bicarbonate solution.
12. Collect the average time for the variables from the other groups.
13. Construct a bar graph of the data (x-axis: variable, y-axis: average time)

Results

Microscope Work

Draw (fill the space) and label a cross section of the C3 leaf. Label both types of mesophyll cells, the air sacs and stomata.

Spinach Leaf Assay 7cm

Trial 1 24°

Your Variable

Leaf Segment	Time (minutes)	
1	2:15	2.28
2	4:25	4.42
3	4:25	4.42
4	5:25	5.42
5	6:20	6.33
6	6:24	6.4
7	6:40	6.67
8	6:43	6.72
9	7:10	7.17
10	7:36	7.6
Average		

Trial 2 11cm

Your Variable: 25°

Leaf Segment	Time (minutes)	
1	7:11	7.18
2	7:18	7.3
3	7:26	7.43
4	7:26	7.48
5	7:26	7.48
6	7:30	7.97
7	8:13	8.22
8	8:13	8.22
9	9:31	9.52
10	9:31	9.52
Average		

X - axis avg time
y - axis diff trials (distance)

- Formula
- Autosum
- Avg.

Table 3: Summary

Variable	Average Time (minutes)

Conclusions

1. Define autotroph and heterotroph.

2. Write the overall general reaction for photosynthesis and state where photosynthesis occurs within the plant cell. chloroplaots

$6CO_2$

3. Explain the experimental design of today's lab (how the rate of rising of the leaf segments is a measure of the rate of photosynthesis) in terms of leaf structure, the overall reaction for photosynthesis, and the solubility of carbon dioxide and oxygen in water.

it took longer for the electrons to excite bc the light was further away (oxygen)

4. How did your alteration of the environment affect the rate of photosynthesis—which condition made photosynthesis go fastest? Slowest? Make sure you support your responses with data. Did your data agree with accepted scientific facts? Why or why not?

Explain what happen

remove oxy
replaced w/ bicarb
when it gain CO2 it
rised again

EXPERIMENT 10: CELLULAR RESPIRATION

Objectives

1. State the overall general reaction for cellular respiration (Relate to the First Law of Thermodynamics) and state the eucaryotic organelle involved.
2. Name and briefly describe the three pathways involved in the complete aerobic oxidation of glucose.
3. Define the following: homeothermic, poikilothermic, and endothermic.
4. State a survival advantage to being homeothermic.
5. Describe how raising the rate of its cellular respiration enables an endothermic animal to raise its body temperature. (Relate to the Second Law of Thermodynamics).
6. State the hypothesis being tested in this experiment and describe how the experimental design of this experiment will enable you to test that hypothesis.
7. Given the necessary data calculate the following for the mouse:
    a. milliliters of oxygen consumed per minute
    b. milliliters of oxygen consumed per minute per gram

Introduction

Cellular Respiration

During cellular respiration, the chemical bond energy within carbohydrates is transformed into the chemical bond energy of the high-energy phosphate bond of adenosine triphosphate—a readily usable form of energy for the cell. Keep in mind that as energy is converted from one form (chemical bond energy of carbohydrates) to another (chemical bond energy of ATP), some energy is always lost as heat as explained by the Second Law of Thermodynamics.

Respiration can occur under aerobic or anaerobic conditions. Although anaerobic respiration plays an important role in many metabolic activities, aerobic respiration is a more efficient energy recovery system. Aerobic respiration yields more energy than does anaerobic respiration and results in the complete breakdown of glucose to carbon dioxide and water. Under aerobic conditions, the many steps in the respiration of the simple 6-carbon sugar, glucose, can be summarized as:

$$C_6H_{12}O_6 + 6O_2 \rightarrow 6CO_2 + 6H_2O + 36\ ATP$$

The complete oxidation of a glucose molecule occurs by three pathways: glycolysis, the Krebs cycle, and the electron transport chain. During glycolysis, the 6-carbon glucose is converted through a series of enzymatic reactions into two 3-carbon molecules of pyruvate. When oxygen is present (aerobic conditions), pyruvate enters the mitochondrion from the cytosol, undergoes another series of enzymatic reactions known as the Krebs Cycle. During the Krebs Cycle, hydrogen atoms are removed from intermediate metabolites and split into positively charged hydrogen ions (protons) and negatively charged high-energy electrons. The electrons are shuttled through a series of oxidation-reduction reactions by way of the electron transport chain found in the cristae of the mitochondria. The energy of these electrons is incorporated into the high-energy phosphate bond of ATP molecules. Finally, at the end of the chain the electrons, together with the hydrogen ions join with oxygen to form water; hence the need for oxygen in aerobic respiration.

There are several ways to measure the rate of cellular respiration in an organism. They include measuring the heat given off, the amount of glucose used, the amount of oxygen consumed, or the amount of carbon dioxide released. This exercise will attempt to indirectly measure respiration by measuring the amount of oxygen consumed.

Respiration in Animals

Homeothermic animals have the ability to maintain a nearly constant internal temperature as opposed to poikilothermic animals whose temperature depends more heavily on their environmental conditions. Most homeothermic animals are endothermic which means they maintain a warm body temperature by means of internal (metabolic) heat production.

In mammals, the hypothalamus, a part of the brain, regulates body temperature by balancing heat production and heat loss. To raise body temperature, the following hypothalamus controlled events occur:

- The adrenal gland releases epinephrine into the blood. This increases the rate of cellular respiration which in turn increases heat production.
- The thyroid gland releases thyroxine. This increases the rate cellular respiration, which increases heat production.
- An increase in muscle tone causes shivering and the energy-requiring muscle contractions involved in shivering increases heat production.
- Vasoconstriction decreases the flow of warm blood from the body core to the skin, reducing heat loss.

To lower body temperature the following events occur:

- The adrenal gland decreases the release of epinephrine into the blood. This decreases the rate of cellular respiration which in turns decreases heat production.
- The thyroid gland decreases the release of thyroxin. This also decreases the rate of cellular respiration and the heat production.
- A decrease in muscle tone prevents shivering and decreases heat production.
- Vasodilatation (an increase in the diameter of blood vessels) increases the flow of warm blood from the core of the body to the skin, causing heat loss through the following mechanisms.
    - Radiation: transfer of heat as infrared heat rays from a warmer object to a colder one without involving physical contact.
    - Conduction: transfer of heat from the surface of the body to any object in contact with the body.
    - Convection: transfer of heat by the movement of a liquid or gas between areas of different temperatures.
    - Evaporation: Perspiration increases and the water evaporates. Since water has a high heat of evaporation the water holds lots of heat as it evaporates.

It is evident that a homeothermic animal, such as a mouse, will increase heat production by increasing its rate of cellular respiration and decrease heat production by decreasing its rate of cellular respiration. From the equation for respiration, it can be concluded that the amount of oxygen taken in and the amount of carbon dioxide released will increase as respiration increases. Keep in mind that inhaled oxygen enters the blood via the lungs and the blood then delivers this oxygen to all the cells of the body where it is used for cellular respiration. The carbon dioxide given off during cellular respiration is picked up from the cells by the blood and carried to the lungs where it is exhaled.

Glycolysis -> transfer reaction -> Krebs cycle -> ETC/ chemi osmosis

$C_6H_{12}O_6 + 6O_2 \rightarrow 6CO_2 + 6H_2O$ + heat + ATP

Materials and Methods

Microscope Work

The slide you will observe is a cluster of cells. The darkened round bodies are the nuclei. Upon careful observation, you will see that these cells are filled with mitochondria. These organelles are about the same size as the bacterial cells you may have observed on your cheek cells. Using high power, draw and label one cell.

Mouse Respiration Experiment

Materials

- safety goggles
- gloves
- metabolism chamber
- mouse
- soap bubble solution
- soda lime
- paper towels
- balance
- tray
- ice

Procedure

1. Fill the cage 3/4 full with paper towels. This will restrict the movement of the mouse.
2. Mass the wire cage using the balance.
3. Have the instructor place a mouse into the wire cage of the metabolic chamber. (Figure 10.1).
4. Use the balance to determine the mass the mouse and metabolism cage. Calculate the mass of just the mouse.
5. Cover the bottom of the metabolism chamber with a thin layer of soda lime, thus absorbing the expired CO_2 and allowing for the measurement of oxygen consumption only. CAUTION: Soda lime is caustic—do not let it touch your skin or the mouse.
6. Clamp a small thermometer to the cage and place the cage into the metabolism chamber. Arrange the cage so the thermometer can be easily read.
7. Wet the inside of the pipette that has been inserted through the rubber stopper. By wetting the inside surface, the chances of the soap bubble seal breaking is reduced. Also wet the outside of the stopper. This will help ensure a tight seal.
8. Insert the rubber stopper into the metabolism chamber, being sure no air leaks occur.
9. Let the mouse acclimate to its new environment for 15 minutes before you begin taking measurements.
10. Record the starting temperature.

11. Place bubble solution at the end of the pipette to form a seal that separates the exterior atmosphere from that found inside the chamber. As the mouse inhales and consumes oxygen, the soap bubble seal should move forward down the pipette towards the mouse.
12. Measure the time, in minutes, required for the soap bubble to travel 5.0 ml along the pipette. If the bubble breaks and has not traveled 5.0 ml, then restart with a new bubble. Do this for three consistent 5.0 ml runs. It may take several trials to get three consistent runs.
13. Record the ending temperature.
14. Calculate the milliliters of oxygen consumed per minute by the mouse. Calculate the ml of oxygen consumed per gram of mouse.
15. Repeat the experiment at a lower temperature by placing the chamber in a tray that contains ice. Allow the temperature inside the chamber to drop to about 16°C and again let the mouse adjust to the new temperature for 15 minutes before you begin. More ice may have to be added to maintain the temperature. After the equilibration period, record the starting temperature, and proceed as above.
16. Have the instructor return the mouse to the cage.
17. Dispose of the soda lime in the labeled waste beaker.

Figure 10.1: Respiration Chamber

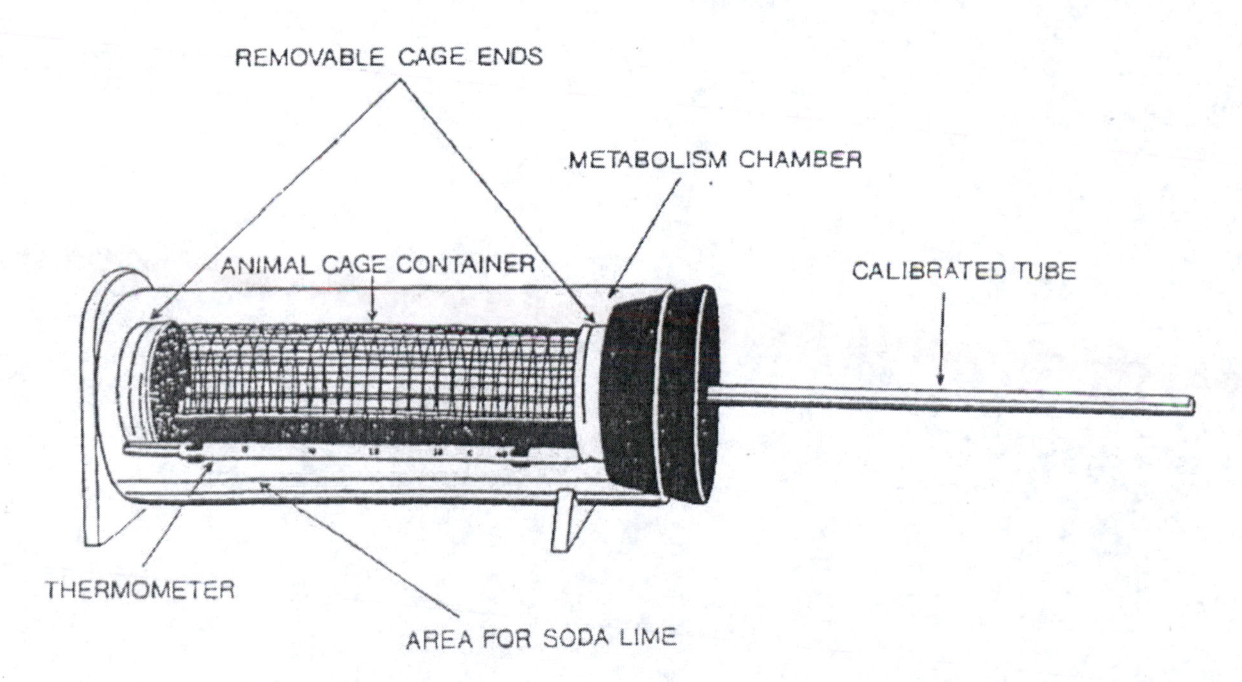

Results

Microscope Work

Draw one cell (fill the space) and label the membrane, mitochondria and nucleus of one cell.

Mouse Respiration Experiment

What are the hypotheses being tested in this lab exercise?

1.

2.

Mass of mouse + cage: 124.4 g 25°

Mass of empty cage: 82.99 g

Mass of mouse: __________ g

Trial 1

Starting Temperature: 25 C°

Ending Temperature: 25 C°

Trial Run	ml of O_2 consumed	Time elapsed (min)
1	5.0	2.14
2	5.0	2.02
3	5.0	2.12
Totals		

General Activity Level of Mouse:

Calculate the ml oxygen consumed per minute as follows:

- Divide the total ml O_2 consumed for the consistent trials by the total minutes.
- Record the ml O_2 consumed/min in the summary table.
- In order to have a metabolic rate that can be compared with an animal of a different mass, we must correct the calculations considering the mouse's mass.
- Perform the following calculation and record this information in the summary table.

ml O_2 consumed per min/mass of the mouse = ml O_2 per minute per gram of mass

Trial 2

Starting Temperature: 16° C°

Ending Temperature: 13 C°

Trial Run	ml of O_2 consumed	Time elapsed (min)
1	5.0	2.17
2	5.0	2.39
3	5.0	2.31
Totals		

General Activity Level of Mouse:

Calculate the ml oxygen consumed per minute and the ml oxygen per minute per gram of mass. Complete the following summary table by obtaining data from the other groups.

Column A	Column B	Column C	Column D	Column E	Column F
		ROOM TEMPERATURE		COLD TEMPERATURE	
Group #	Mass (g)	ml O_2 per min	ml O_2 per minute per gram of mass	ml O_2 per min	ml O_2 per minute per gram of mass
Calculations			Col. C/Col. B		Col. E/B
1	45.4	2.49	0.055	8.47	0.19
2	60.0	1.93	0.032	3.54	0.059
3	41.5	2.32	0.06	2.01	0.05
4	.576	2.69	0.046	2.92	0.089

Construct a bar graph that compares the mass of the mice to the ml oxygen per minute per gram of mass at both temperatures.

Conclusions

1. Give the overall general reaction for cellular respiration. State what eukaryotic cell organelle is involved.

2. Define homeothermic and endothermic.

3. What effect did lowering the temperature have on the mouse's oxygen consumption? What is the effect on cellular respiration? Is this result what you would expect? (Look at the graph.) Explain and be sure to support your response with data.

4. If you used a poikilothermic animal, such as a frog, for your experiment, what would you have expected to happen to the respiration rate when the environment's temperature was dropped?

5. How should activity level affect oxygen consumption?

6. According to the data collected (Look at the graph.) on the mice, is there a correlation between the mass of the mouse and its room temperature respiration rate. Explain and support with data.

7. When you exercise, your body gets hotter. Stored chemical energy in the form of glucose and muscle glycogen is burned to provide your muscles with ATP so they can move. How is this an example of the Second Law of Thermodynamics?

8. Describe how cellular respiration and photosynthesis are related.

9. Do plant cells conduct cellular respiration? Why or why not?

EXPERIMENT 11:
DNA: GEL ELECTROPHORESIS

Objectives

1. Explain how PCR makes copies of DNA.
2. Discuss the role of *Taq* in PCR.
3. Describe how DNA fragments can be separated by an electrophoresis chamber.
4. List the functions of the following materials that are used in electrophoresis: agarose gel, methylene blue, buffer solution and gel box.
5. Explain the significance of short tandum repeats (STRs).
6. Be able to read a DNA fingerprint and interpret their meaning.

Introduction

PCR

Polymerase chain reaction (PCR) combines *in vitro* DNA synthesis with specific annealing (binding) of complimentary DNA sequences. It requires a solution containing the template (the original DNA), the DNA polymerase enzyme, buffer to optimize the conditions for the polymerase, Mg^{+2} ions for enzyme progression, two primers specific for the DNA region of interest, and free nucleotides to build the new DNA strands.

Short single-stranded segments of DNA whose nucleotide sequences are complimentary to the DNA regions flanking the region of interest on the template DNA are used as primers for the DNA synthesis. One primer will bind (anneal) one strand of the template DNA and the second primer will bind to the other strand. Primers are designed to have a sequence specific for the DNA region of interest; consequently they bind only to that region when the DNA is at a high temperature. To facilitate annealing and discourage non-specific bonding, the primers are 15 to 25 base pairs in length and the PCR reaction is performed using temperatures just below the maximum temperature for annealing. The temperature at which primers bind depends on the number of guanine-cytosine and adenine-thymine bonds between the primer and the template DNA. Custom primers of any sequence can be purchased from several biotechnology supply companies.

To perform the PCR reaction a large excess of both primers are used in relation to the amount of template DNA. To start the reaction the double-stranded DNA is denatured into single strands by heating the sample solution to 95ºC (Figure 11.1). (In a cell, helicase and stabilizing proteins accomplish this step.) The mixture is then cooled so the primers can anneal (bind) to the complementary regions (Figure 11.2). (In a cell primase and RNA nucleotides are used to begin the DNA polymer.) Because the primers are in great excess, a primer molecule will bind to each template strand of DNA before the template strands can re-anneal to each other. The DNA polymerase enzyme then uses the free nucleotides to synthesize a second DNA strand from each of the original template strands (Figure 11.3). The annealed primer serves as the starting point. Because DNA polymerase can only add nucleotides to the 3' end of the end of the DNA, the synthesis of the new DNA proceeds in one direction (Figure11.4). The result is now two long, double-stranded DNA (Figure 11.5).

The process is called polymerase chain reaction because it is repeated over and over in a "chain reaction" or a cycle. To repeat the process the DNA is heated again to denature it, the primers are allowed to anneal, and then the polymerase synthesizes DNA to make the double-stranded products. With each cycle, the number of DNA strands doubles. Therefore, if the process starts with one double-stranded piece of DNA, after one cycle there will be two double-stranded pieces of DNA. In the second cycle when the DNA is denatured there will be four single-stranded templates, and at the end of the second cycle there will be four double-stranded pieces of DNA (Figure 11.6). In the third cycle there will be eight single-stranded templates, and at the end of the third cycle there will be eight double-stranded pieces of DNA. Typically a PCR repeats for 30 cycles. (Figure 11.7) therefore in a typical PCR each original DNA molecule will yield 2^{30} molecules (1,073,741,824 molecules of DNA). (Figure 11.8)

When the PCR technique was first developed, fresh polymerase enzyme had to be added for each cycle because the heat denatured the protein enzyme in addition to the double-stranded DNA. After the isolation of the bacterium, *Thermus aquaticus (Taq),* from the hot springs in Yellowstone National Park, the use of PCR increased dramatically. Since the *Taq* bacterium adapted to the extremely hot environments, its DNA polymerase can withstand high heat without denaturing. When *Taq* polymerase is used in PCR, fresh polymerase does not need to be added to each cycle. Instruments called thermocyclers were developed which accurately and automatically cycle through the temperature changes for the PCR. Kary Mullis received the Nobel Prize in 1993 for his development of the PCR in the early 1980's.

PCR is so efficient that enough product DNA is produced that it is visible on an agarose gel, even if only minute quantities of DNA were available to begin the reaction. This technique is used to produce many copies of a specific DNA region for cloning, sequencing, forensics, paternity testing, detention of diseases or other analysis.

Forensics

Amplify the DNA
polymerase chain reaction (PCR)

DNA fingerprint

highly variable regions of DNA
(STR) short tandom repeat

① 2 single strand DNA primers
② taq polymerase (heat stable)

PCR in Pictures

Melting

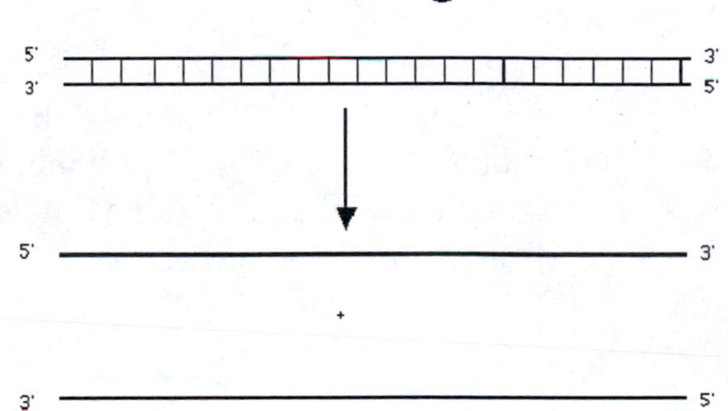

Figure 11.1: Samples are heated to 94-96°C for one to several minutes to denature (separate into single strands) the target DNA

Anneal primers

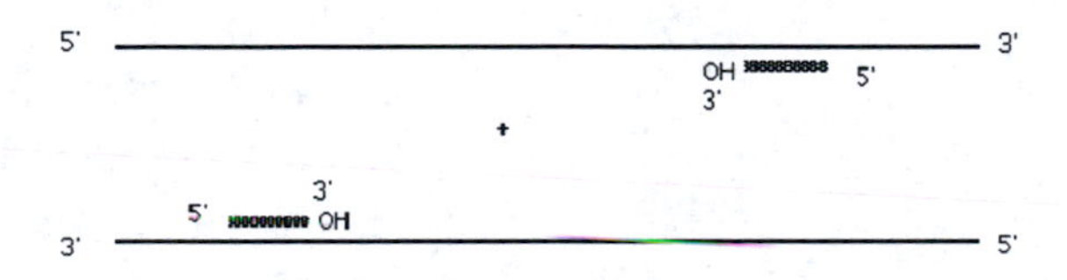

Primers are short (18-24 nt), synthetic oligonucleotides.

Sequence chosen to complement target strands.

Distance between 5' termini determines size of PCR product.

Figure 11.2: The temperature is lowered to 50-65°C for one to several minutes allowing the left and right primers to anneal (basepair) to their complementary sequences on the DNA template.

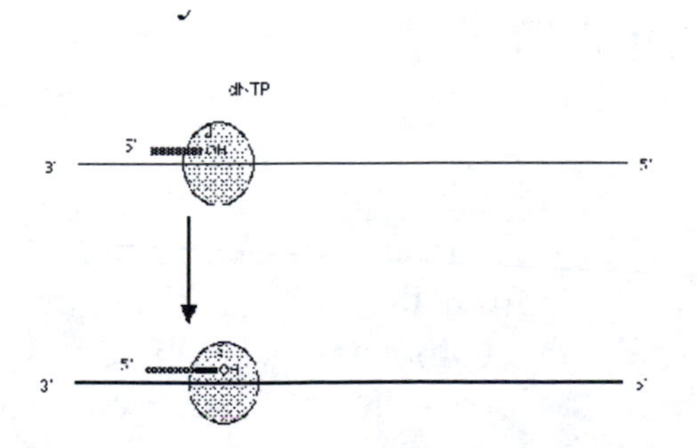

Figure 11.3: The temperature is raised for one to several minutes, allowing *Taq* polymerase to attach at each priming site (where primers have annealed) and extend (synthesize) a new DNA strand.

Polymerase Extension

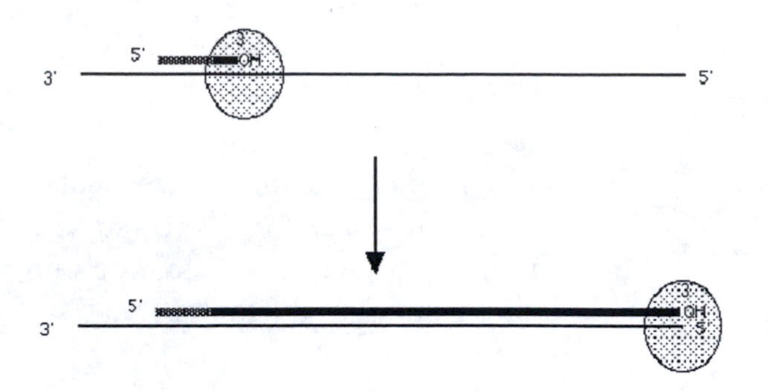

Figure 11.4: Only one of the original DNA strands is being shown. The *Taq* polymerase continues the extension of the new, complementary strand.

Products from one strand

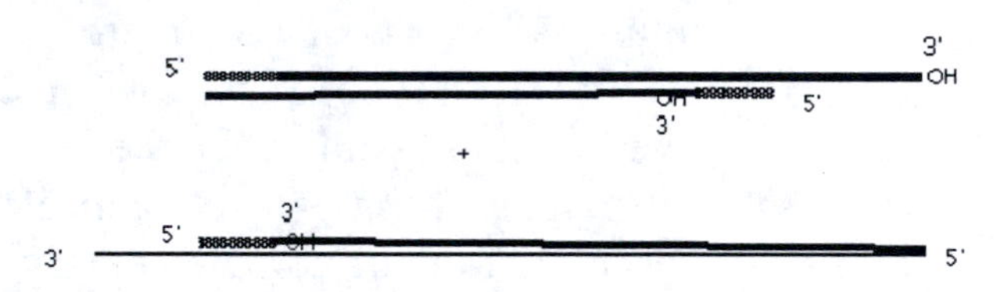

"Parental" DNA strand (bottom) was template for new strand beginning at 5' end of primer (top).

Melt and anneal new primers.

Figure 11.5: After 1 complete cycle, 2 strands of DNA are produced; however, neither is of the area that you are interested in. The mixture is reheated so that the DNA will become single stranded and the primers will be able to anneal to the each of the resulting single strands that are produced.

Products of second cycle of PCR

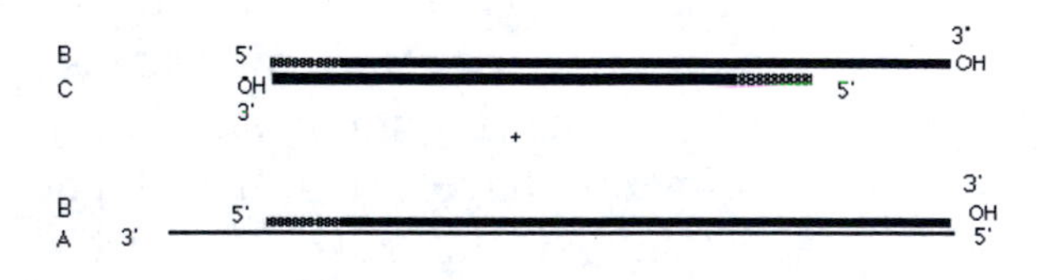

Parental strand (A) remains.

First product strand (B) copied from parental.

Second product strand (C) copied from first product (B).

Figure 11.6: The diagram tracks the results of one strand of DNA from the original template. After 2 complete cycles there will be 4, long DNA strands created but none will be of the area of interest.

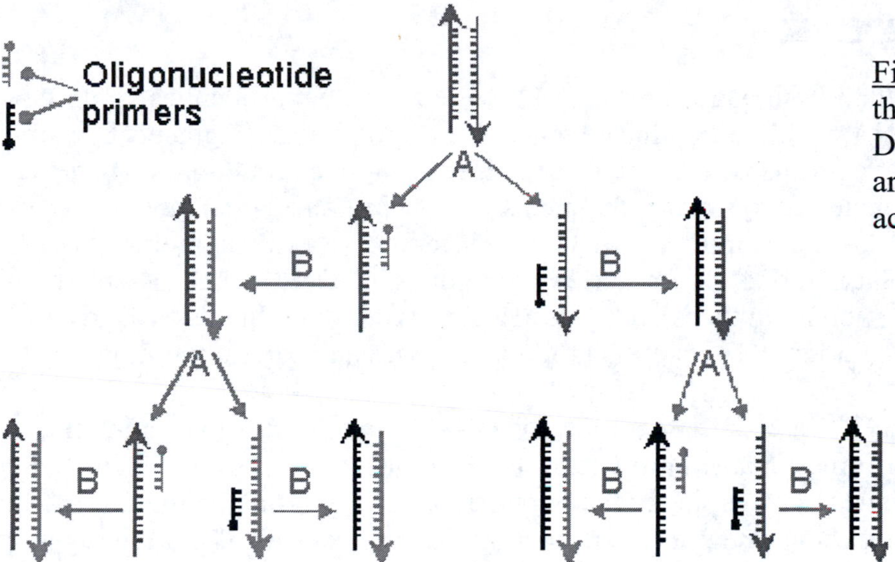

Figure 11.7: After the 3rd cycle there will be eight strands of DNA, two of which are of the area of interest. Each strand now acts as a template.

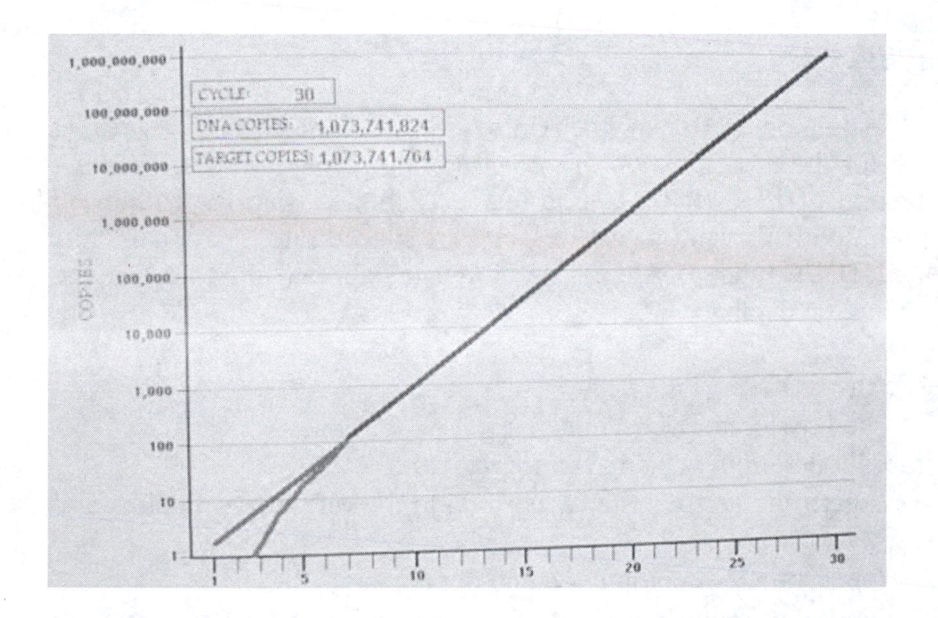

Figure 11.8: The amount of DNA that is produced grows geometrically because of the number of DNA strands increases. After 30 cycles the resulting DNA will be 99.9% target DNA (the gene of interest.)

PCR & DNA Fingerprinting

Humans have about 3 billion base pairs (bp) of DNA with 99.9% being identical from one person to another. The DNA could be examined by cutting it into various fragments by using restriction enzymes. The fragments could then be separated using electrophoresis. However, the 3 billion bps would generate too many fragments, producing an overlapping smear of bands in a gel. Fortunately, the 3 million bp (0.1%) differences between individuals are concentrated in certain places along chromosomes. Variations in DNA sequences between individuals are termed polymorphisms and sequences with the highest degree of polymorphism are very useful for DNA analysis in forensics cases and paternity testing.

Short Tandem Repeats (STRs) are short sequences of DNA, normally 2 to 5 bps in length, that are repeated numerous times in an end-to-end, or tandem, manner. For example the 16 bp sequence of "gatagatagatagata" is 4 copies of the tetramer "gata". PCR can be used to detect these STRs. Although the chromosomal locations and the base sequences of the repeats at a given site are the same from person to person, the number of the repeats at a given location is highly variable; thus there are many alleles for a given region. This means that the DNA fragments that result from PCR are different sizes because the area of DNA being amplified as a different number of repeats (bases). Since an individual may have different numbers of repeats on the maternal and paternal chromosomes, you may obtain one or two DNA fragments of different sizes from PCR using a single primer pair.

A National DNA Databank

The Federal Bureau of Investigation (FBI) of the United States has been a leader in developing DNA technology for the identification of perpetrators of violent crime. In 1997, the FBI announced the selection of 13 STR tetrameric loci that show variations within the population. DNA profiles on individuals using these 13 loci were placed into a national DNA database, CODIS (Combined DNA Index System). All forensic laboratories that use the CODIS system can contribute to the database.

There are many advantages to the CODIS STR system:

- It has been widely adopted by forensic DNA analysts.
- STR alleles can be rapidly determined using commercial kits.
- The STR alleles are discrete and are inherited according to the rules of Mendelian and population genetics.
- The data is digital and thus suited for computer databases.
- The database is growing thus contributing to the analysis of STR frequency in different human populations.
- STR profiles can be determined with minute amounts of DNA.

Gel Electrophoresis

Electrophoresis is a technique used in the laboratory that results in the separation of charged molecules; in this case the DNA fragments that were produced from the PCR. DNA is a negatively charged molecule, and is moved by electric current through a matrix of agarose, a chain of sugar molecules that is extracted from seaweed. Purified agarose in its powdered form is insoluble in a buffer at room temperature; but it dissolves in boiling water. When it starts to cool, it undergoes what is known as polymerization. Rather than staying dissolved in the water or coming out of solution, the sugar polymers crosslink with each other, causing the solution to "gel" into a semi-solid matrix much like "Jello" only firmer. The more agarose that is dissolved in the boiling water, the firmer the gel. While the solution is still hot, it is poured it into a mold called a "casting tray" so as it polymerizes it will assume the desired shape. The molecules are now in a web-like matrix.

DNA samples are placed into wells at one end of the gel and the gel is then placed into an electrophoresis chamber that contains running buffer (which supplies ions for the electric current and maintains the pH). An electric current will then run through the gel, pulling the DNA towards the positive pole. Smaller fragments will move through the gel faster than larger fragments; thus smaller fragments will move further in the gel.

After the electric current is turned off, the DNA will be "stuck" in a specific location in the gel. To visualize the DNA, various staining techniques can be used to visualize the DNA (the bands). By comparing the bands of suspects with the CODIS data base or children with potential parents, positive identifications can be made.

Autoradiographs

The DNA in the bands are made radioactive (created by either by using radioactive probes that will complementary bind to specific areas in the bands or by using radioactive nucleotides in the PCR procedure). By doing a Southern Blot procedure, the DNA is transferred from the (fragile) gel to a (sturdy) nitrocellulose paper. The nitrocellulose paper is placed in contact with X-ray film. This exposed film is called an autoradiograph and will be used to interpret the DNA.

Our Experiment

In this simplified activity, you will be looking at the PCR results from one region in the DNA. In actuality, you would need to looks at the PCR results from multiple regions in the chromosomes to definitely identify an individual. There might be a 1 in 20 chance that two individuals would randomly match in one particular location and a 1 in 30 chance that two individuals would randomly match in another locations. Therefore there is a 20 x 30, or 1 in 600 chance that these two individuals would match in both locations. Increasing the number of locations being examined increases the likelihood of a positive identification.

The Scenario

Jim and his girlfriend, Sarah, stopped on their way home from a late movie to make a late-night withdrawal from an ATM in an isolated parking lot. Sarah stayed in the car while Jim walked up to the machine. As soon as his punched in his PIN, a man in a ski mask and gloves came out of the bushes and swung a tire iron at Jim. As Sarah frantically dialed 911 on her cell phone, Jim was brutally beaten by his attacker. Jim fought back, clawing at the man, but Jim was unarmed. The attacker withdrew cash from the ATM and fled. As he ran away, Sarah saw him pull the mask from his face, revealing a distinctive Roman nose silhouetted against the parking lot lights.

The police and an ambulance arrived quickly and took Jim to the hospital where he eventually recovered from his injuries. Based on the distinctive nose, Sarah was able to identify two likely suspects from the mug shots at the police station. The police were quickly able to take the suspects into custody, but both provided an alibi. Sarah told the police how Jim had clawed at his attacker. One suspect had a scratch on his face which he claimed was from shaving. The other suspect had scratches on his hands which he claimed were from some work he had done on his car. The police dispatched the forensics experts to the hospital where they recovered samples of blood and skin from underneath Jim's fingernails.

The lab technician received the samples of blood and skin that were collected as evidence. The samples were tiny but she was able to extract a small amount of DNA from them. The technician was also supplied with epithelial (cheek) cells from the two suspects. PCR was conducted on the following samples:

- Suspect 1
- Suspect 2
- Victim's (Jim) DNA
- Evidence (to be sure that the evidence was not contaminated with Jim's DNA)

It is your job to conduct the electrophoresis on the samples and to analyze the results.

(1) Denature DNA 95°
(2) primers bind (annealing) 55°
(3) extension 72°
DNA doubles
1 molecule → 1 billion
electrophoresis

Materials and Methods

Gel Electrophoresis

Materials

- safety goggles
- 0.7% pre-made agarose gel
    - Instructor: Add stain to the cooled but still liquid solution
        - Concentrated (dropper bottle) Caroline Blu™: 8 drops/100.0 ml

          OR
        - Final Stain solution: 0.7 ml/100.0 ml
- electrophoresis buffer
    - Instructor: Check the concentration of the stock solution and dilute accordingly
        - Add stain to the running buffer
            - Concentrated (dropper bottle) Caroline Blu™: 24 drops/500.0 ml

              OR
            - Final Stain solution: 3.1 ml/500.0 ml
- 20 µl micropipette set to 15 µl
- electrophoresis chamber
- DNA samples
- staining container
- Carolina Blu™ final stain

Procedure

1. The electrophoresis chamber has been prepared with running buffer and up to two gels. The power source is on the right and the wells should be away from you.
2. Using a 20µl micropipettor, transfer the entire DNA sample from the tube into the appropriate well in the gel. Use a new tip for each sample.
    a. Far left well: Blue tube, suspect 1
    b. Next well: Green tube, suspect 2
    c. Next well: Pink/red tube, Jim's DNA
    d. Next well: White tube, evidence
3. After the wells have been loaded, place the cover on the electrophoresis chamber, plug in the unit and use the Main Switch (in the back of the unit) to turn on the unit.
4. Press the Power Button and set the timer for 30 minutes. (Figure 11.9)
5. Set the voltage to 100V.
6. Press the Run/Pause/Stop button. The red LED light indicates that the system power is on.

Figure 11.9: Power Supply LCD

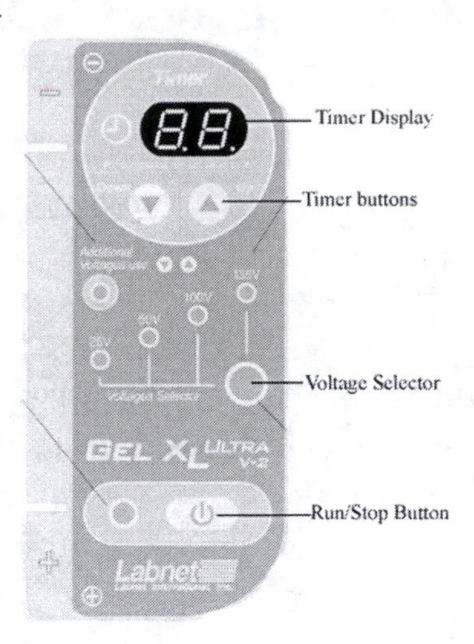

7. To confirm proper operation of the power supply, look for bubbles rising from the electrodes and that dye movement is in the proper direction.
8. At the end of 30 minutes, the power automatically turns off. Use the Main Switch to turn the machine off. Unplug the unit. Remove the gels from the box and place them on a light box to see the bands. If the bands are not visible, place the gels into a container and pour approximately 10 ml of Caroline Blu™ final stain onto the gel.
9. After approximately 4 minutes, return to stain to the bottle for reuse.
10. Place the gel back on the light box. If the bands are still not clearly visible put them into a plastic bag with a small amount of the running buffer. Add 2 drops of the concentrate stain and leave until them until the next lab period.
11. Sketch the gels. Compare the evidence DNA to the DNA from the two suspects.

Part 2: Reading Autoradiographs

Autorad #1

This film contains the evidence from five actual paternity cases. Ease case consists of the DNA from the mother, the child, the alleged father and a mixture of DNA from the child and the alleged father. Quality control lanes include a fragment of DNA of a known size (K562 and GSE-132) and sizing markers (DNA fragments with many known sized fragments). You will be assigned one of the five paternity cases to read and interpret.

Autorad #2

This film depicts known DNA drawn from a sexual assault victim and two suspects. DNA evidence from the crime scene was separated into sperm cells and female epithelia cells. The victim's husband was also tested. As in the 1st autorad, quality control DNA was also used. You will need to read and interpret this film.

Results

Gel Electrophoresis

Sketch the fragmentation patterns (fill the space) that appeared. Make sure you label the contents of the wells. Place the following: +, -, large fragments, small fragments at the appropriate sides of the gel.

Reading Autoradiographs

Autorad #1
For the case you were assigned, is the alleged father the biological father?

Case Number: __________

Autorad #2
Do the female cells that were found at the crime scene belong to the victim?

Where did the sperm come from? Suspect 1, suspect 2 or the husband

Conclusions

1. Briefly describe how DNA fragments can be separated by gel electrophoresis. What property of DNA is used during the running of a gel?

2. For the gel that you ran, which suspect is guilty?

3. Was the evidence DNA contaminated with Jim's DNA?

4. Who are the only individuals possessing the same DNA fingerprints (profiles)?

5. Why is PCR an extremely valuable tool for the molecular biologist?

EXPERIMENT 12:
DNA EXTRACTION

Objectives

1. Explain the necessary steps to do a crude extraction DNA from a eukaryotic cell.
2. Discuss the physical and chemical properties of DNA, such as solubility and high molecular mass.

Introduction

The extraction of DNA from any type of cell involves (a) disruption of all membranes (b) removal of proteins and other cellular debris from the nucleic acids and (c) a final purification. The purity of the desired DNA sample, will dictate which procedure will be used.

Since membranes are primarily phospholipids and proteins, the membranes of many bacterial and human cells can be dissolved with a detergent. The detergent will also denature many of the cell's proteins, especially the Dnases. If these enzymes are not denatured, the DNA from a lysed cell will be broken into small pieces by these enzymes. Any enzymes not denatured by the detergent will be denatured by heat.

If a highly purified DNA sample is required, it is important to remove all remaining proteins from the sample. This is accomplished by treating the samples with proteases (protein destroying enzymes) and/or extractions with the organic solvent, phenol. (Proteins are soluble in phenol and DNA is not.) Further purification procedures, such as precipitation, dialysis and high-speed centrifugations, may also be necessary.

In the lab, you will do a crude extraction of DNA from dried peas. Peas, salt and water are placed into a blender to separate the cells from one another. (The role of the salt is to neutralize the charge of the DNA's sugar phosphate backbone. This makes the DNA less hydrophilic) The detergent helps to break down the plasma membrane. The meat tenderizer breaks down proteins. The alcohol is less dense than the water so it will float on the top of your mixture. DNA normally can dissolve in water but will not dissolve in alcohol. The addition of the salt allows the DNA molecules to stick together.

Materials and Methods

Materials

- safety goggles
- 84.0 g of dried split peas
- 1.0g table salt
- 200.0 ml cold water
- 30.0 ml liquid dish detergent
- meat tenderizer
- cold 70-95% alcohol
- blender
- large beaker
- strainer
- test tubes
- test tube holders
- 5 ml pipettes
- flat toothpick
- wooden stick

Procedure

*Steps to be done by the instructor.

1. *Place peas, salt and water into the blender and blend for about 15 seconds.
2. *Pour the "pea soup" through the strainer and collect the liquid in the beaker.
3. *Add the detergent and let sit for at least 10 minutes.
4. Place 3 ml of the mixture into a test tube.
5. Add 1 toothpick full of meat tenderizer. Gently mix by tapping the test tube so that the solution is swirled.
6. Slowly trickle 3.0 ml of cold alcohol down the inside of the tilted tube to form a layer that floats on the top of the cell mixture.
7. Push the wooden stick through the alcohol and cell mixture layer a number of times to make the DNA visible.

Results

Describe (or draw and label) the appearance of the DNA.

Conclusions

1. What is the purpose of the following in a DNA extraction procedure: detergent, tenderizer, salt, and alcohol?

2. Why does the alcohol need to stay on top of the lysate mixture?

3. What information is stored in the DNA molecule?

EXPERIMENT 13: EUKARYOTIC GENE EXPRESSION A COMPUTER SIMULATION

Objectives

1. Describe the purpose of the following parts of a eukaryotic gene: promoter and terminator regions, introns and exons, and start and stop codons. Be able to locate these regions within DNA and/or mRNA sequence(s).
2. Identify the start codon.
3. Define transcription and translation.
4. Given a DNA template segment, be able to determine the mRNA and amino acid sequences that result from the DNA.
5. Define mutation and give examples of various types of mutations.
6. Describe the effects of mutations on gene expression.
7. Construct a eukaryotic gene that results in a polypeptide chain.

Introduction

"Understanding nature's mute but elegant language of living cells is the quest of modern molecular biology. From an alphabet of only four letters representing the chemical subunits of DNA, emerges a syntax of life processes whose most complex expression is man. The unraveling and use of this 'alphabet' to form new 'words and phrases' is a central focus of the field of molecular biology. The staggering volume of molecular data and its cryptic and subtle patterns have led to an absolute requirement for computerized databases and analysis tools. The challenge is in finding new approaches to deal with the volume and complexity of data, and in providing researchers with better access to analysis and computing tools in order to advance understanding of our genetic legacy and its role in health and disease."

National Center for Biotechnology Information

The process by which information encoded in the DNA molecule (the sequences of the nucleotides) is converted into a protein (a sequence of amino acids) is a two step process. During transcription, a RNA molecule which is complementary to the template strand of the DNA molecule, is synthesized. Messenger RNA (mRNA) contains the specific coding sequence for a specific polypeptide chain. The mRNA has the same nucleotide sequence (except uracil has replaced thymine) as the other DNA strand (coding strand). In translation, the nucleotide sequences on the mRNA are converted to amino acid sequences. Three consecutive nucleotides on the mRNA are called a codon which will be converted into either one amino acid, a start or stop signal.

Although the process of protein synthesis and the meaning of the genetic code is universal, eukaryotic genes and mRNA are more complex than ones found in prokaryotic cells. Both types of cells have promoter and terminator regions which control transcription. The promoter region is where the RNA polymerase attaches to begin transcription while the terminator region releases the enzyme. All cells also have start and stop codons which control translation. Although promoter and terminator regions (the sequences of DNA nucleotides) are different in different organisms, these regions when generated by the Gene Explorer Program (The program is available for free at http://intro.bio.umb.edu/GX/) will be as follows:

Promoter

5' – TATAA – 3'

3' – ATATT – 5'

Transcription begins with the first base-pair to the right of the promoter sequence and will proceed to the right. The sequences prior to the start sequence are called Upstream Leader Sequences. Translation will begin with the start sequence.

Terminator

5' – GGGGG – 3'

3' – CCCCC – 5'

Transcription ends with the first base-pair to the left of the terminator sequence. Within the gene there will be a stop sequence and generally other nucleotides will follow the stop (Downstream Trailing Sequences) before the terminator is found. The Downstream Trailing Sequences are transcribed but not translated.

Therefore, a gene generated in Gene Explorer will look like this:

5' – TATAAXXXXXXXXXXXXXXXXXXXXXXGGGGG – 3'

3' – ATATTXXXXXXXXXXXXXXXXXXXXXXCCCCC – 5'

In addition, eukaryotic genes have additional features that are not found in prokaryotic genes. Eukaryotic transcription produces a molecule called pre-mRNA because the molecule is not ready for translation. This molecule is then processed:

- A process called mRNA splicing occurs where introns (a non-protein coding sequence found in the DNA and pre-mRNA) are removed. The remaining sequences, exons (a protein coding sequence found in the DNA and mRNA), are joined.
    - Gene Explorer introns will start with 5'–GUGCG–3'
    - Introns will end with 5'–CAAAG–3'
- Modified guanine nucleotides are added to the 5' end of the mRNA. This is called the cap and enables the ribosome to bind to the mRNA.
    - Gene Explorer will not show the cap.
- As many as 400 adenines can be added to the end of 3' end of the mRNA. This is called the tail. Both the cap and tail extend the "life span" of the mRNA.
    - The program will add 13 As to the end of the mRNA.
        - Remember that there is no complementary sequence to caps and tails in the DNA.

Using the Gene Explorer program you will:

- find promoter (in green) and terminator (in red) regions
- read DNA sequences to produce pre-mRNA
- distinguish intron (non-colored) and exon (various colors) regions
- find start and stop codons (underlined)
- translate mRNA to amino acids
- mutate the DNA and determine the effects on the polypeptide chain

When there is a change in the sequence of the bases on the DNA a mutation has occurred. Although mutation rates are high, cells have mechanisms to repair certain types of mutations; but if the DNA is not corrected the change will be passed on to future generation of cells. Most mutations are considered silent mutations because they will have no discernible effect on the cell because the:

- mutation occurred in an intron region
- cell does not use the protein where the mutation occurred
- resulting amino acid was the same as the original (silent mutation)
- resulting amino acid is very similar to the original
- change did not change the shape of the protein

There are a number of different types of mutations.

- Base Substitution (point mutation): a change in only one pair of nucleotides. The original pair is AT and it is changed to GC.
    - Missense Mutation: results in a different amino acid
    - Silent Mutation: results in the same amino acid in the polypeptide chain
    - Nonsense Mutation: results in a stop codon
- Frameshift Mutation: one or more bases are added or removed

Originally DNA sequences were analyzed by hand but now scientists use complex computer programs (similar to Gene Explorer) to do this labor intensive work. This merger of computers and biological information is called Bioinformatics. To aid scientists in their studies, the government created the National Center for Biotechnology Information (NCBI: http://www.ncbi.nlm.nih.gov). The site contains data bases that contain DNA and protein sequences, tools for their visualization and comparative analysis, and the Online Mendelian Inheritance in Man (OMIM) database, a catalog of human genes and disorders.

Materials and Methods

Materials

- computer with Gene Explorer
- printer (If doing the Design Your Own Gene section)

Table 13.1: Important DNA Sequences

Keep these sequences available as you mutate the DNA. The sequences will help you determine why the amino acid sequences are altered.

GENE REGION	DNA CODING SEQUENCE 5' to 3'
Promoter	TATAA
Terminator	GGGGG
Intron Beginning	GTGCG
Intron End	CAAAG
Start	ATG
Stops	TGA, TAA, TAG

Procedure

Part I: Familiarize Yourself with the Gene Explorer Program

1. From the Desk Top, open the Gene Explorer Program. (If asked, click on the Run Program.) Make sure the program fills the screen. (See Figure 13.2) Start and stop codons are underlined.

Figure 13.2: Gene Explorer Screen

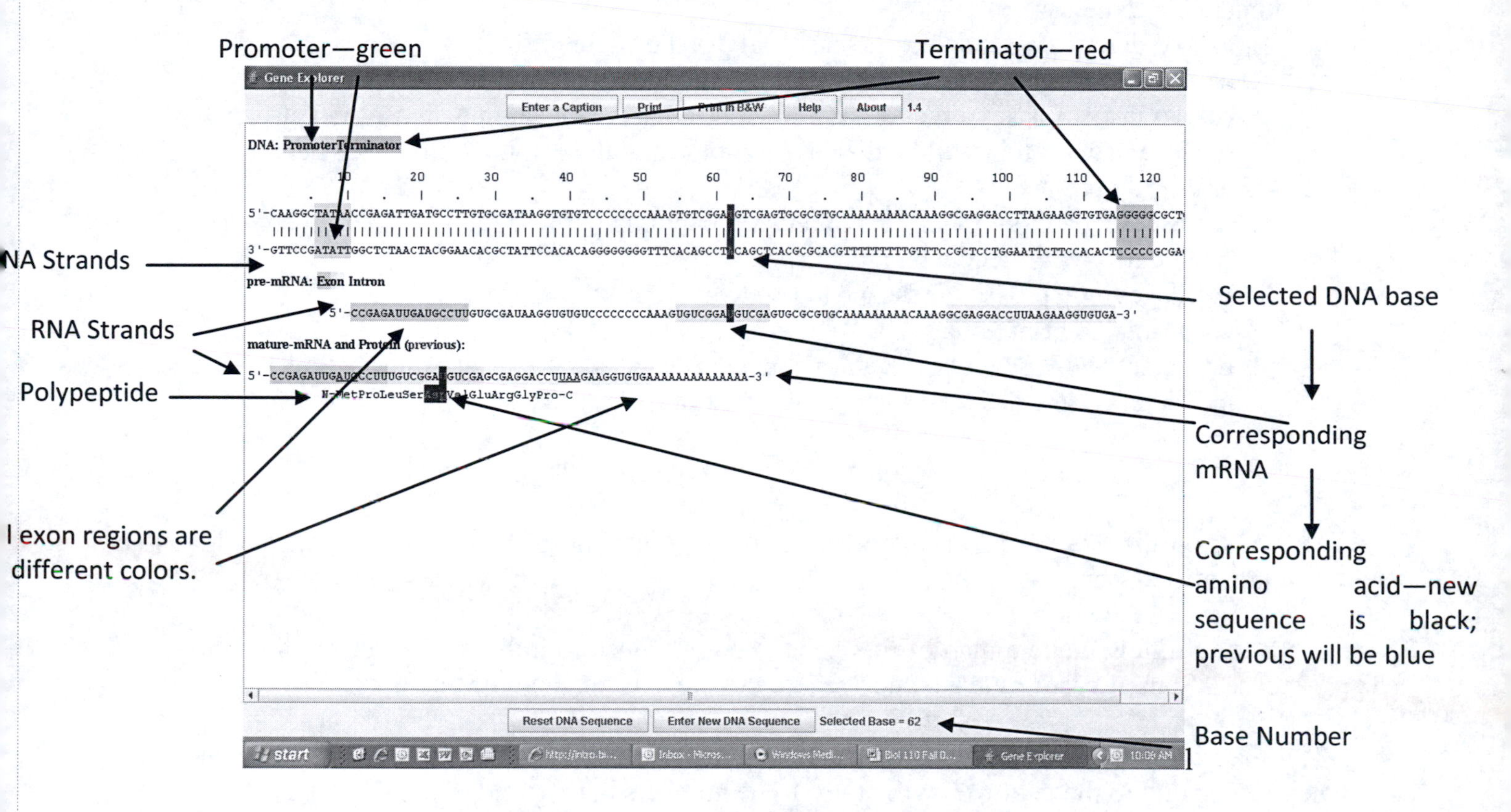

2. DNA is "read" from the 3' to 5' end so in Gene Explorer the bottom strand (template strand) will be transcribed. The complementary, coding strand, is where the sequences in Table 13.1 are located. The resulting mRNA will then be translated from the 5' to the 3' end. In the resulting polypeptide chain, the "N" stands for the unbound amino group and the "C" is the carboxyl group.

3. By selecting a DNA nucleotide, the program will show the corresponding base in the mRNAs and amino acid in the protein. To select a DNA nucleotide:
   a. You can only make a selection in the top DNA strand.
      i. The selected bases and amino acids are hi-lited in dark blue.
   b. Place the pointer at the base you are interested in.
   c. It sometimes is difficult to select the exact base. Try the following:
      i. Use the "→" key to move one base to the right and the "←" key to move one base to the left.
   d. The number of the selected base is shown at the bottom right of the screen.
4. You can edit the DNA (The resulting mRNA strands and protein sequences are automatically updated.) strand by:
   a. To delete a base, select the base and use the "delete" key
   b. To replace the selected base with another base, type lower case a, t, c or g
   c. To insert a base to the left of the selected base, type an upper case A, T, C or G
   d. The new protein will be displayed in blue and the original will be black.
5. Other useful buttons:
   a. Enter a caption: Click this to add a few words to the bottom of the screen. Use this to identify your gene when it is printed.
   b. Page Set Up: Use landscape to print you gene.
   c. Print: To print your gene.
   d. Reset DNA Sequence: Returns the DNA sequence to the starting (default) sequence
   e. Enter a New DNA Sequence: To enter your sequence.

Part II: The Eukaryotic Gene

1. Click on the Reset DNA sequence to return the display to the default sequence. This is an unmodified gene.

2. You will now make a map of this gene. You will need to individually select all of the DNA bases. Start at base 1 and use the "→" key to move through all of the bases.

The map format is illustrated in Figure 13.3. (The diagram is for illustration purposes and does not correspond to the gene you will study.) The numbers correspond to the DNA base.

Figure 13.3: Gene Map

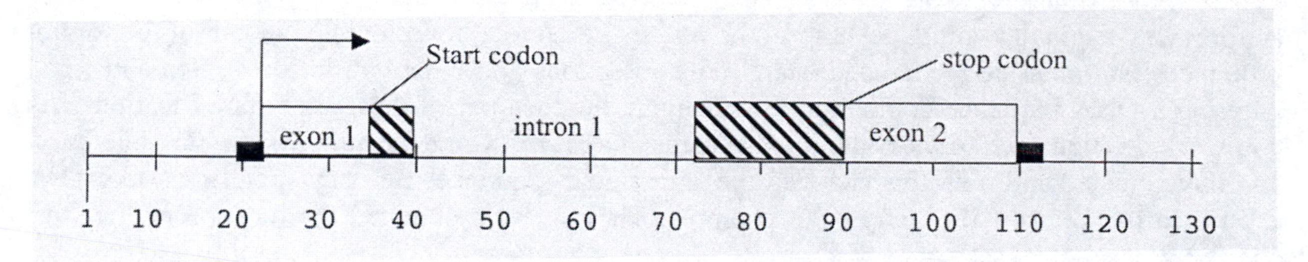

The map coordinates (numbers) are approximate.

- Promoter: small black rectangle at position 20
- Start of transcription: indicated by the bent arrow at position 22
- Exon 1: labeled box and runs from 22 to 40
- Intron 1: labeled bland area and runs from 41 to 72
- Exon 2: labeled box and runs from 73 to 110
- End of transcription: end of the last exon at 110
- Terminator: small black rectangle at position 110
- Start Codon: start of the hashed region in exon 1 at 35
- Stop Codon: end of the hashed region in exon 2 at 90
- Coding Region: the regions that encode the protein are the hashed parts of the exon regions, it extends from 35 to 90 but will not include the intron.

Make a map of the gene that is shown in Gene Explorer. Be sure to include all the features that are listed above. If there are more than 2 exons and 1 intron, be sure to include them.

Results will be placed into the Table found in the Result/Conclusion Section. Pay careful attention to the location in the gene where the mutations will be made. Be very specific about how the mutation altered the amino acid sequence.

3. Complete Worksheet 13.1. This worksheet shows your gene map in table form and will be useful for determining how the mutation altered the amino acid sequence.

4. The poly A tail at the end nor the poly G cap (Remember that Gene Explorer did not add the cap.) at the beginning of the mRNA do not have a corresponding sequence in the DNA. Also note the start codon is not at the beginning of the mRNA molecule. Prior to the start codon are RNA nucleotides that are referred to as the leader sequence which contains binding sites for the ribosomes. The code for the polypeptide begins with the start codon and ends at the stop codon

Part III: Mutations

Before you begin it would be helpful if you make a note to remind yourself of important sequences that must be recognized: start and stop codons and where an intron begins and ends. Altering a DNA sequence could result in changing the location of these important locations. As you work through each of the mutation problems, make sure you determine what type of deletion you have created and how this changed the amino acid sequence. **Be very specific as to HOW the mutation altered the gene.** (For example: The mutation changed the location of the start codon.)

Mutation 1

- Change the C/G base-pair at position 58 to a T/A base-pair.
  - Select base 58—check the position number at the bottom of the screen.
  - Type ‘t’ (It must be a lower case.)

Mutation 2

- Reset the DNA sequence.
- Delete the T/A base-pair in position 26.
  - Select base 26—check its position.
  - Hit the delete key.

Mutation 3

- Reset the DNA sequence.
- Change the A/T base-pair in position 51 to T/A base-pair.
  - Select base 51—check its position.
  - Type “t” (lower case).

Mutation 4

- Reset the DNA sequence.
- Change the T/A base-pair in position 21 to a G/C base-pair.
  - Select base 21—checks its position.
  - Type “g” (lower case).

tt - frameship

In the previous exercises, you were instructed to make specific changes to the DNA and then determine the type of mutations that were made. In the next series of mutations, you will need to create a specific type of mutation. Using the Genetic Dictionary (Figure 13.4) you will need to find an appropriate DNA base-pair to change or delete. Remember the coding in the dictionary is for codons (mRNA) and you will be changing DNA nucleotides. You will making changes to the 5' to 3' (top) strand. The coding on this strand will be same as the resulting mRNA.

Figure 13.4: The Genetic Dictionary for mRNA Codons

Table of mRNA codons

First Base	Second Base: U	C	A	G	Third Base
U	phenylalanine phenylalanine leucine leucine	serine serine serine serine	tyrosine tyrosine STOP STOP	cysteine cysteine STOP tryptophan	U C A G
C	leucine leucine leucine leucine	proline proline proline proline	histidine histidine glutamine glutamine	arginine arginine arginine arginine	U C A G
A	isoleucine isoleucine isoleucine START methionine	threonine threonine threonine threonine	asparagine asparagine lysine lysine	serine serine arginine arginine	U C A G
G	valine valine valine valine	alanine alanine alanine alanine	aspartate aspartate glutamate glutamate	glycine glycine glycine glycine	U C A G

Mutation 5

- Reset the DNA sequence.
- Produce a nonsense mutation.
  - On the result sheet, indicate the nucleotide number that was changed and the new base-pair.

Mutation 6

- Reset the DNA sequence.
- Produce a silent mutation in an exon region.
  - On the result sheet, indicate the nucleotide number that was changed, the original base-pair and the new base-pair.

Mutation 7

- Reset the DNA sequence.
- Produce a mutation that produces no mRNA and therefore no protein.
  - On the result sheet, indicate the nucleotide number that was changed, the original base-pair and the new base-pair.

Part IV: Design Your Own Gene (Optional: Instructor's Discretion)

Design an entirely new gene that you have invented. Using Gene Explorer, your gene must include the following:

- Sequences prior to the promoter region
- At least two exon regions
- At least one intron region
- Upstream Leader and Downstream Trailing Sequences
- At least 5 amino acids and there codes are distributed within all the exons

When you have completed your gene, click on the Add a Caption and type in your name. Although you will be doing the work as a group, each student must submit their own copy of the gene with their own name on the paper. Use the print button to print a copy of your gene. Construct a gene map of your gene.

Tips

- Use the "Enter New DNA Sequence" button and delete all the data
- Refer to Figure 12.1 for important sequences that need to be included in your gene.
- Type in a small, random sequence of bases.
- You will be constructing the 5' to 3' strand (the top one).
- Type in the promoter sequence
- Add the leader sequence
- Add the start
- Add sequences for the amino acids
- Add the stop
- Check to see that the protein is being constructed
- Once you have the protein return to the New DNA Sequence and using the back button find the location where you will add the intron sequence
- Make sure your gene works.

Results and Conclusions

The Eukaryotic Gene

Make a map of the gene that is shown in Gene Explorer.

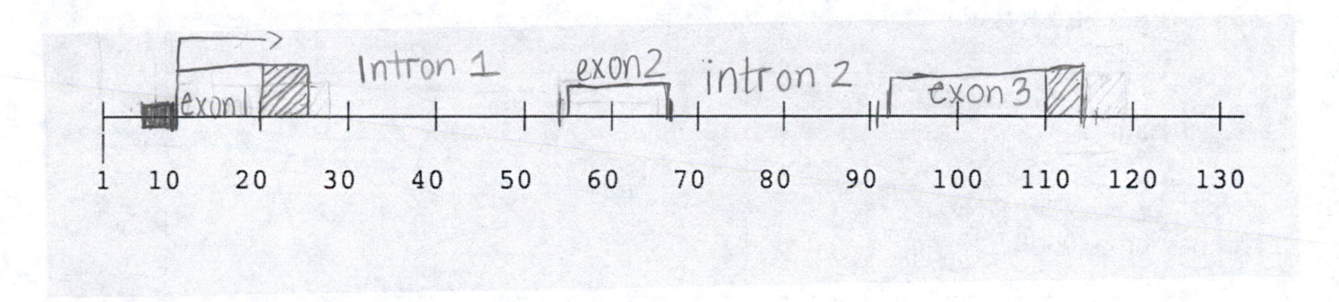

Place the amino acid sequence in the chart that is on the next page.

Complete the following chart.

Region	Length	Multiples of 3?
Intron 1		
Intron 2		
Exon 1 (from start)		
Exon 2		
Exon 3 (to stop)		
Exon Total		

Worksheet 13.1: Complete the following chart and use this information to help you determine how the mutation altered the amino acid sequence. (Use Table 13.1 to help locate the regions and remember you are using the coding (top) strand of the DNA.

REGION	BASE NUMBERS
Promotor Region	6-10
Exon 1	11-20
Start Codon	21-24
Intron 1 Beginning Sequences	25-54
Intron 1 Ending Sequences	
Exon 2	
Intron 2 Beginning Sequences	
Intron 2 Ending Sequences	
Exon 3	
Stop Codon	
Terminator Region	

Mutations

Original amino acid sequence										and how many more	What region of the original DNA strand did the mutation occur?	**Specifically**, how did the mutation change the amino acid sequence?	Mutation type
Mutation 1	met	pro	leu	leu	ASP	val	glu	Arg	Gly	Pro	exon2	regroup insert new codon	
Mutation 2	met	pro	cys	arg	met	ser	ser	ser	Asp	+3 (+4)	exon2	frame shift	
Mutation 3	met	pro	cys	glu	Asp	leu	leu	lys	val	0	intron1	less amino acids	
Mutation 4	met	ser	ser	glu	Asp	leu	leu	lys		0	exon1		

Original amino acid sequence											In what region of the original DNA strand does the mutation occur?	How did the mutation change the amino acid sequence?
Mutation 5 1.Nucleotide # changed: 2.Original bp: 3. New bp:	met	pro	leu	ser	asp	val	glu				exon 3	become stop codon
Mutation 6 1.Nucleotide # changed: 2.Original bp: 3. New bp:	met	pro	leu	ser	asp	val	gly	val	gly	pro	exon 1	
Mutation 7 1.Nucleotide # changed: 2.Original bp: 3. New bp:											promoter	no protein made

Design Your Own Gene: (Optional: Instructor's Discretion)

Gene Map

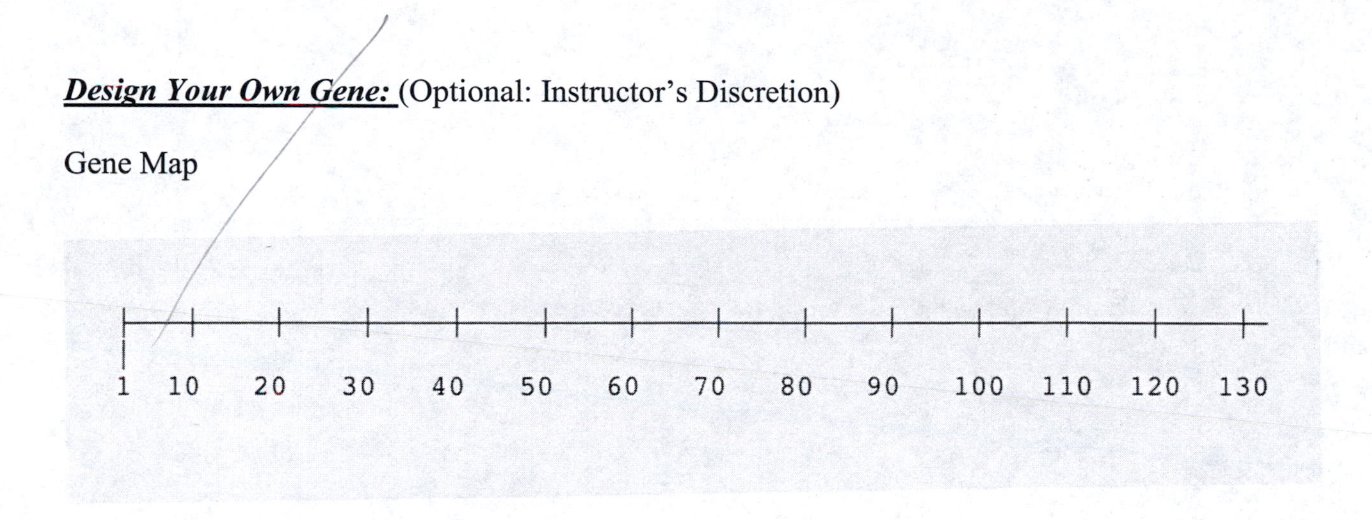

Attach the print out of your gene.

Answer the following questions

1. Why can introns be of any length?
2. Why are the individual exons not in multiples of 3 but yet their total length is a multiple of 3?
3. Below is a mRNA that contains a cap and tail. What is the resulting amino acid sequence?

5'GGGGGGGCGAAGAUUGAUGACUAGAAAUUGGGAUCCUACGUGAGAAGGUGUG
AAAAAAAAAAAAAAA3'

EXPERIMENT 14:
MITOSIS

Objectives

1. Differentiate the different stages of mitosis.
2. Draw the stages of mitosis as they appear under the microscope.
3. Determine which stage a cell spends the most through the least amount of time in.

Introduction

Eukaryotic cells that divide undergo a cell cycle: an orderly sequence of events that extends from the formation of a new cell to when that cell divides. This type of division will lead to identical cells where each daughter cell is identical to each other and they are both identical to the original mother cell. The Cell Cycle can be broken down into the following steps:

Interphase: the non-dividing cell

a. G1: cell increases its organelles and size
b. S: DNA is replicated
c. G2: final synthesis of proteins, especially those needed for division

M Phase: cell division

a. Prophase: chromatin material shortens to become chromosomes; nuclear membrane disappears
b. Metaphase: chromosomes line up on equatorial plate
c. Anaphase: chromatids move to the poles
d. Telophase: the nucleus reappears and cytokinesis occurs

In this activity, you will examine root tips (mitotically active cells) of an onion to determine the relative length of time of each of 5 discernible steps of the cell cycle. One basic assumption of this exercise is that the more cells there are in a particular stage, the longer is the duration of that stage.

Figure 14.1: Mitosis in Plant Cells

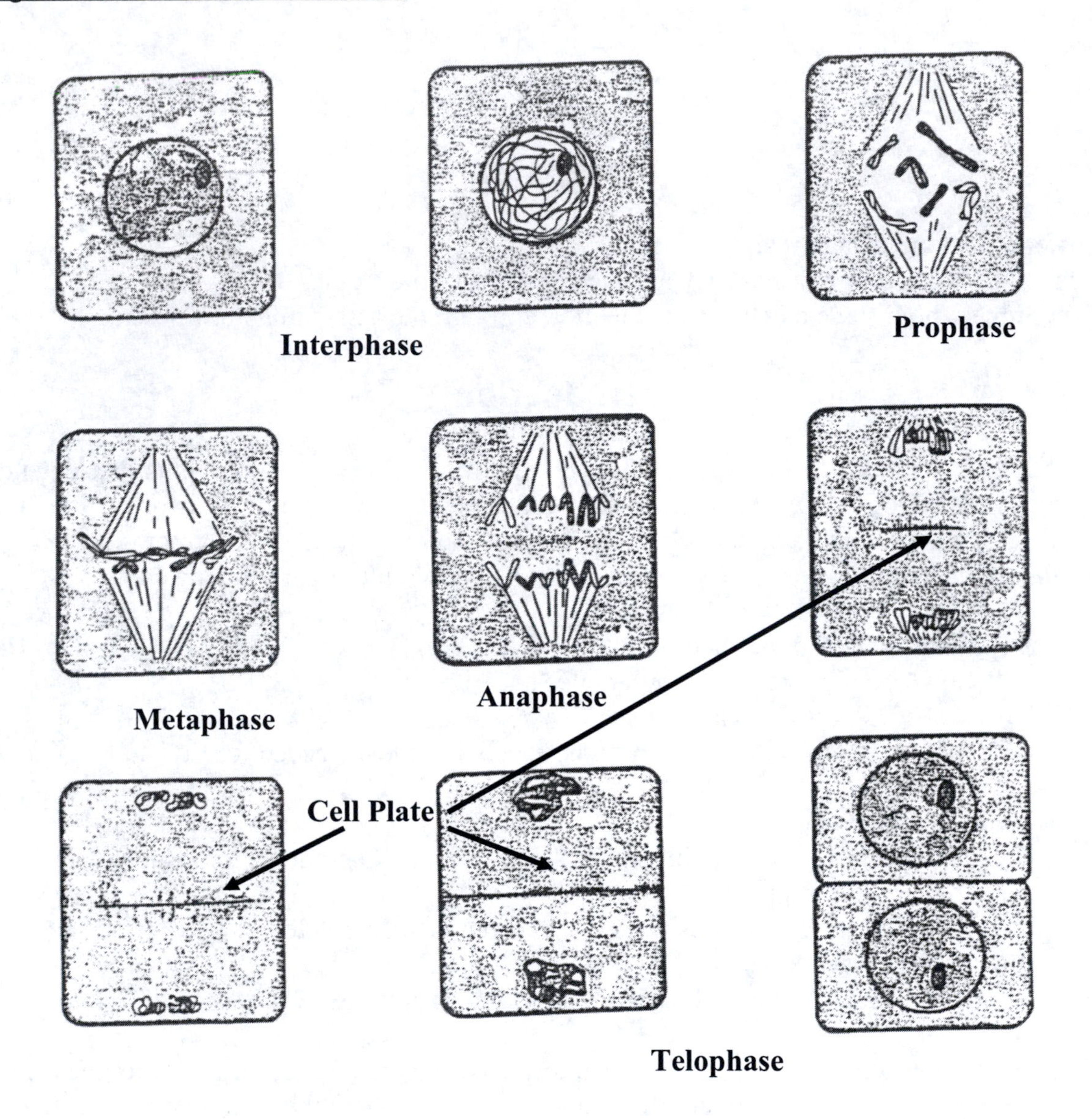

Materials and Methods

Materials

- microscope
- onion (*Allium*) root tip slide

Procedure

1. Obtain a root tip slide. Using low power, scan the root tip and locate a region just behind the rounded tip that contains a high number of cells undergoing mitosis. This is called the Zone of Division.

   Figure 14.2: Tip of Onion Root

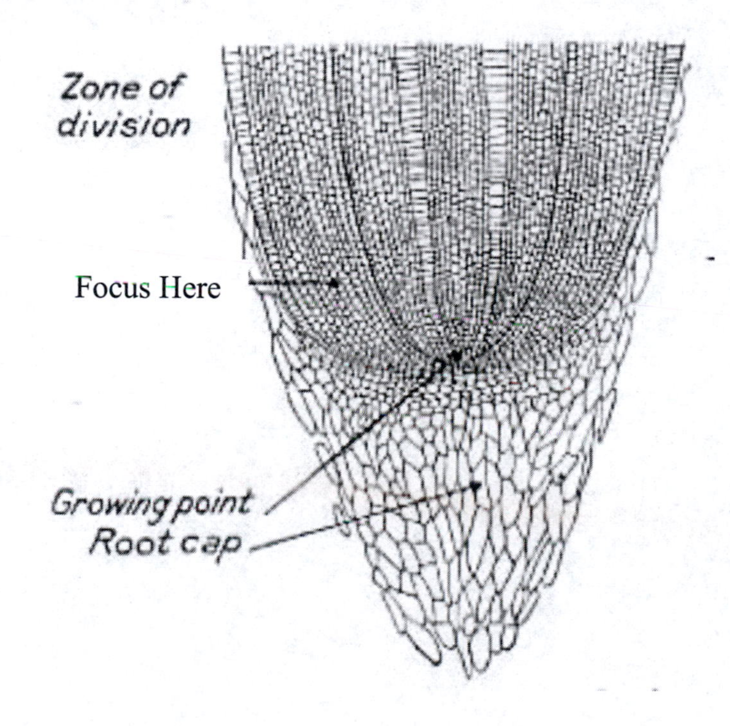

2. Change to high power.
3. Determine which stage the cell spends the most time in by determining which stage is most prevalent on the slide.
4. Determine which stage the cell spends the next amount of time in by estimating which stage is the 2nd most abundant on the slide.
5. Continue until you have determined an order for the remaining stages.

Results

1. What is the order of time spent in interphase, prophase, metaphase, anaphase and telophase?

    a. (longest)

    b.

    c.

    d.

    e. (shortest)

Conclusions

1. Sketch the five visible stages of the animal cell cycle. Your cell has a diploid number of 4.

EXPERIMENT 15: MENDELIAN GENETICS: A COMPUTER SIMULATION

Objectives

1. State Mendel's Laws of Segregation and Independent Assortment.
2. Define the following: gene, allele, dominant, recessive, genotype, phenotype, homozygous, heterozygous, and incomplete dominance.
3. Use a Punnett Square or probability to predict genotypes and phenotypes in a monohybrid, dihybrid and incomplete dominant crosses.
4. When give a pure breeding organism for a given trait be able to determine the dominant and recessive trait.
5. Know the expected F_2 phenotype ratios for a dihybrid cross.
6. When given the chi-square formula and table, perform a chi-square test on the F_2 generation of a monohybrid, dihybrid, and incomplete dominance crosses and explain the significance of the results.

Introduction

Mendelian Genetics derives its names from the work of the Austrian monk, Gregor Mendel. Mendel began his studies of heredity in 1857 using the common garden pea. He choose the pea plant because they were readily available, easy to grow, grew rapidly and the distinctive traits bred true and reappeared each year. The plants normally self-fertilize because the pistil and stamens in the flowers are entirely enclosed by the petals. By carefully removing the male stamens, fertilization between different plants can be achieved.

For his initial experiments, Mendel chose plants that showed the two variations (alleles or genes) for a single characteristic (gene). Since only one characteristic was followed at a time these crosses are called monohybrid crosses. For example, Mendel crossed pure breeding tall plants with pure breeding short plants. This P_1 (parental) Generation produced an F_1 (first filial) Generation that was completely tall. The other characteristic, shortness, appeared to be lost in this generation. However, when these F_1 plants were allowed to self-pollinate, the next generation, F_2 had both tall and short plants. Furthermore, Mendel noted a consistent 3:1 ratio of tall to short plants. Thus the short trait was not lost in the F_1 generation but was masked in some fashion such that it did not express itself. The trait capable of being masked is referred to as the recessive trait and the other trait is the dominant trait.

From his monohybrid experiments, Mendel formulated his first law of inheritance: The Law of Segregation. In modern terms, the law states that the various traits of an organism are controlled by genes and these genes are transmitted from one generation to the next in a predictable way. Each gene can exist in two alternative forms, alleles. These alleles segregate during meiosis, so that each gamete contains only one allele. During fertilization, the new organism receives one allele from each parent, thus restoring the two alleles per gene.

In Mendelian Genetics, it is customary to identify the alleles by letter. The dominant allele is expressed by an uppercase letter. Thus in following the trait for the height of the plant, the dominant, tall allele is indicated by a “T”. The recessive, short allele must be the lowercase of the dominant letter, thus is a “t”. The genotype of an organism is its genetic make-up (allele combination) and the outward appearance of the organism is the phenotype. When an organism is true-breeding, it contains identical alleles for a trait and is said to be homozygous. Therefore, the P_1 tall plant is TT and the P_1 short plant is tt.

The genotype of all the F_1 tall plants are Tt because the tall parent can only pass along a T and the short parent can only pass along a t. Since this generation contains two different alleles for the dominant phenotype, the genotype is heterozygous. (Geneticists now understand that each allele is located on one of homologous chromosomes of a eukaryotic cell. When called upon, each gene will be transcribed and for a heterozygous individual two different protein structures will be made: a normal, functioning protein and an abnormal, non-functioning one. When compared to the homozygous dominant, the heterozygote is making half the amount of normal protein, but this amount is sufficient so that the heterozygote appears dominant.) The genotypes of the F_2 generation can be predicted by constructing a Punnett Square where the possible gametes of the parents are placed on the outside of the square and the possible offspring will be inside the square.

Figure 15.1: Punnet Square of an F_1 Generation

Gametes:↓ Male, → Female	T	t
T	TT tall	Tt tall
t	Tt tall	tt short

The phenotypic predications for the F_2 generation are 3 (dominant tall):1 (recessive short). As long as the plant contains the tall gene, that plant will be tall. The genotypic predications for this generation are 1 (homozygous dominant):2 (heterozygous)1: (homozygous recessive).

Mendel conducted monohybrid crosses on seven different characteristics and determined the dominant trait for each characteristic. He then wanted to determine if the inheritance of one trait influenced the inheritance of another trait. These types of crosses are called di-hybrid. In one cross Mendel wanted to know if the height of the plant (tall or short) was connected with the color of the plant’s flowers, purple or white.

Again Mendel started with plants that were pure breeding tall-purple or short-white. These P_1 plants were crossed and the F_1 plants were all tall-purple. Since the P_1 genotypes were TTPP (homozygous dominant for both characteristics) and ttpp (homozygous recessive for both characteristics) the F_1 generation must all be TtPp. Again, Mendel allowed the F_1 plants to self-fertilize and noticed that noted that 9/16 of them were tall with purple flowers, 3/16 were tall with white flowers, 3/16 were short with purple flowers and 1/16 were short with white flowers. However, when the individual characteristics were examined ¾ were tall and ¼ were short and ¾ were purple and ¼ were white.

This information gave us Mendel's second law of inheritance, The Law of Independent Assortment which said that members of any gene pair segregate from one another independently of the members of the other gene pairs. This occurs in a regular way that ensures that each gamete contains one allele for each locus (site), but the alleles of different loci are assorted at random with respect to each other in the gametes. We now know that this law will only occur if the genes are located on different chromosomes or if they are the same chromosome they are placed far enough apart so as to not be influenced by crossing over.

Mendel studied traits where one trait was completely dominant to a recessive. If Mendel had used a plant such as the ornamental Sweet Pea, he would have seen another type of inheritance pattern that we call Incomplete Dominance. In this case a pure breeding red is crossed to a pure breeding white and all pink flowers result. When two of these F_1 pinks are bred, the result is 1 red, 2 pinks and 1 white. In this case, the red gene is incompletely dominant to the white and a third phenotype results. (In this case, the amount of protein produced by the heterozygote is insufficient to produce enough pigmentation in the petals; thus the flowers are pink.) Prior to Mendel's work people believed in a Blending Theory of Inheritance where each parent contributes equally to what their offspring look like. At first glance the production of pink flowers supports this idea; however when two pink plants are bred, the resulting F_2 generation shows a phenotypic ratio of 1 red: 2 pink: 1 white. If a Blending Theory was to be supported this generation should be all pink.

Data Interpretation

A comparison of the traits will be expressed as ratios and are expressed as the relative number of the expected phenotypes (# counted/# of the smallest population). In Mendel's experiments on the height of the plants, in the F_2 generation he counted 303 tall plants and 96 short plants. The ratio of tall to short is:

$$\frac{303}{96} : \frac{96}{96} = 3.2{:}\,1$$

303 tall
96 short

There are times when your counted numbers look close to a theoretical value, as they appear in our example. However, it must be determined if your values are close enough to the theoretical values, so a chi-square analysis needs to be done.

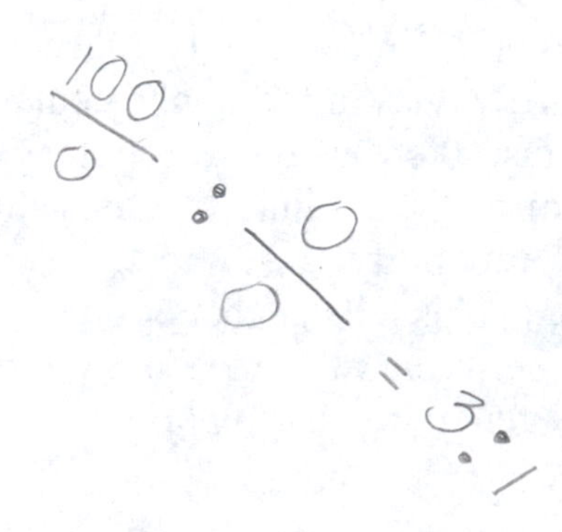

The chi-square formula is:

$$\chi 2 = \frac{(o_1 - e_1)^2}{e_1} + \frac{(o_2 - e_2)^2}{e_2}$$

Figure 15.2: Chi-Square Table

Not Significant								Significant	
d.f. P	.95	.90	.85	.70	.50	.30	.10	.05	.01
1	.004	.016	.064	.148	.455	1.07	2.71	3.84	6.64
2	.103	.211	.446	.713	1.38	2.41	4.60	5.99	9.20
3	.352	.584	1.00	1.42	2.37	3.66	6.25	7.82	11.3

Using the above example, we need to determine if our results are indeed close enough to a theoretical value to say that the observed plants are an F_2 generation.

Figure 15.3: Calculation of Expected Values for the F_2 Generation

	Observed	Expected
Tall (1)	303	399*.75 = 299
Short (2)	96	399 – 299 = 100
TOTALS	399	399

The above numbers are then plugged into the formula.

$$\chi^2 = \frac{(303 - 299)^2}{299} + \frac{(96 - 100)^2}{100}$$

$$\chi^2 = \frac{(4)^2}{299} + \frac{(-4)^2}{100}$$

$$\chi^2 = \frac{16}{299} + \frac{16}{100} = .05 + .16 = .21$$

This value (.21) must now be placed into the Chi Square Table. The degrees of freedom (df) (number of values in the final calculation of a statistic that are free to vary) in this experiment is 1 because there were two choices (classes) for the outcome—tall or short. (The df is the number of choices minus 1). Our value of .21 falls between .148 and .455 and this correlates to a P (probability) of .7 to .5. This value falls under the major column heading of not significant. This tells us that our results are not significantly different than the expected results (We are between 50% and 70% confident in our results.) thus our hypothesis that the seeds are an F_2 generation is supported.

Materials and Methods

Adapted from Pea Plant Genetics

Materials

- PC using Windows
- Pea Plant Genetics Software

Figure 15.4: Pea Plant Genetics Breeding Control Panel

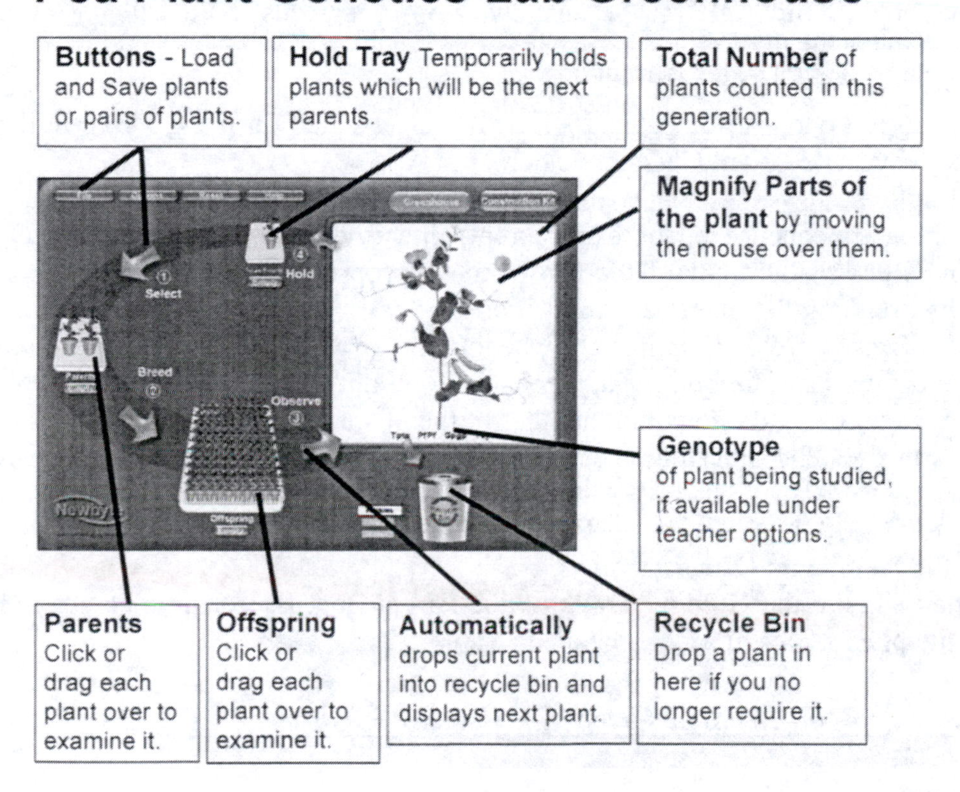

Experiment 1: Monohybrid Cross for Pea Color

Introduction

In this experiment we will examine the inheritance of a single gene which governs the color of the edible pea in pea plants. The peas will either be yellow or green. By repeating Mendel's crosses you will determine which color is dominant. You will also follow this trait in three generations of plants.

Procedure

1. From the Desk Top, double click on the Pea Plant Genetics Click anywhere on the opening screen. Click OK on the next two pop-up screens. (We will not be saving the data.)
2. Empty the Parents and Offspring Trays.
3. Click on File, Load, Saved Parents. (The files are located on the C Drive in the Program Files, Newbyte, Pea Plant Genetics folder.
4. Select Exp05.pep.
5. To examine the phenotype of each parent, drag the first parent into the Observation Platform. Record the color of the pea in the appropriate place in Table 1.1. Return the parent by dragging the plant back to the Parents Tray. Repeat for the other parent.
6. Click the Breed arrow. Examine the phenotypes of all the offspring by clicking on the Observe arrow. Use the bottom of Table 1.1 to count the offspring. Place a plant with a number in the 30s and one with a number in the 80s into the Holding Tray by clicking on the Hold arrow. The other plants will be placed into the Compost Bin (the remaining 98 plants).
7. Complete Table 1.1.
8. Empty the Parent and Offspring areas.
9. Drag the plants in the Holding Tray into the Parent Box. Breed these plants. Observe and record the phenotypes of all the plants in Table 1.2.

Experiment 2: Monohybrid Cross for Flower Position

Introduction

In this experiment we will examine the inheritance of a single gene which governs the position of the flower on the plant. The flower's position will either be axial (toward the main stem) or terminal (at the top of the plant). By repeating Mendel's crosses you will determine which position is dominant. You will also follow this trait in three generations of plants.

Procedure

1. Click on the Reset.
2. Follow the steps from the previous experiments, but select Exp07.pep.
3. Repeat Steps 5 to 8 from Experiment 1 and complete Tables 2.1 and 2.2.

Experiment 3: Dihybrid Cross for Pea Color and Flower Position (Optional: Instructor's Discretion)

Introduction

In your previous experiments you determined the dominant genes for each of these traits. You will now determine if the color of the pea influences the position of the flower in the pea plant.

Procedure

1. Click on the Construction Kit button at the upper right.
2. Click the Reset button to ensure you have a common pear plant.
3. Change the Pod Colour to Yellow. Click on Use as Parent #1, then OK.
4. Change to Flower Position to Terminal and the Pod Colour to Green. Click on Use as Parent #2, then OK.
5. Click on Greenhouse and repeat Steps 5 to 8 from Experiment 1 and complete Tables 3.1 and 3.2.

Experiment 4: Flower Color in Sweet Peas (Optional: Instructor's Discretion)

Introduction

In this experiment you will determine the inheritance pattern for flower color in the ornamental sweet pea plant.

Procedure

1. Click on the Reset.
2. Follow the steps from the previous experiments, but select Exp19.pep.
3. Repeat Steps 5 to 8 from Experiment 1 and complete Tables 4.1 and 4.2.

Experiment 4: Flower Color in Sweet Peas

Table 4.1: F_1 Cross of Flower Color in Sweet Peas

<table>
<tr><td colspan="3">Parent Descriptions:
1st Parent: white
2nd Parent: red</td></tr>
<tr><td>Red Flowers</td><td>White Flowers</td><td>Pink Flowers</td></tr>
<tr><td></td><td></td><td></td></tr>
<tr><td>Total Red Flowers:</td><td>Total White Flowers:</td><td>Total Pink Flowers:</td></tr>
</table>

Table 4.2: F_1 Cross Results for Sweet Pea Flower Color:

<table>
<tr><td>Red Flowers</td><td>White Flowers</td><td>Pink Flowers</td></tr>
<tr><td></td><td></td><td></td></tr>
<tr><td>Total Red Flowers:</td><td>Total White Flowers:</td><td>Total Pink Flowers:</td></tr>
<tr><td colspan="3">Ratios: (red:pink:white)</td></tr>
</table>

Table 4.3: Chi-Square Calculations

Phenotype	Observed (o)	Expected (e)	o - e	$(o-e)^2$
Total				

CONCL[illegible]ONS

Experiment 1

1. What pea color is dominant? Recessive? How c[illegible] arrive at this conclusion?

2. Using "C" and "c", what are the genotypes of th[illegible]l F_1 plants?

3. What ratio of yellow and green peas would you ex[illegible] the F_2 generation?

4. By doing a Chi Square analysis on the F_2 data, ca[illegible] confirm that these plants are a Mendelian F_2 Generation? Show your work.

Experiment 2

1. What flower position is dominant? Recessive? How did y[illegible]ve at this conclusion?

2. Using "P" and "p", what are the genotypes of the P_1 and F_1 p[illegible]

3. What ratio of axial and terminal flowers would you expect in t[illegible]neration?

4. By doing a Chi Square analysis on the F_2 data, can you confi[illegible] these plants are a Mendelian F_2 Generation? Show your work.

5. In the natural setting, the trait that is most common is called the w[illegible]e and the rarer type is the mutant. With respect to Mendel's peas, green peas and axia[illegible]r position are the wild-types. By examining your data from Experiments 1 and 2:
    a. Which allele is dominant (expressed)?
    b. Which allele is the wild type?
    c. A mutation results in a phenotype that is different from t[illegible]ype. Dominant means that if you have inherited that allele you have tha[illegible]e. Do mutations always cause an dominant allele to become a recessiv[illegible] Support your reason with data.

Experiment 3

1. Do the genes for pea color and flower position exhibit Independent Assortment? How do you know this?

2. By doing a Chi Square analysis on the F_2 data, can you confirm that these plants are a Mendelian F_2 Generation showing Independent Assortment? Show your work

Experiment 4

1. What type of inheritance pattern is being exhibited in this experiment? How do you know this?

2. What are the genotypes of all the phenotypes?

3. By doing a Chi Square analysis on the F_2 data, can you confirm that these plants are an F_2 Generation showing Independent Assortment? (Hint: There are 3 choices for flower color.) Show your work.

Experiment 4: Flower Color in Sweet Peas

Table 4.1: F_1 Cross of Flower Color in Sweet Peas

Parent Descriptions: 1st Parent: white 2nd Parent: red		
Red Flowers	White Flowers	Pink Flowers
		𝍸 𝍸 𝍸 𝍸 𝍸 𝍸 𝍸 𝍸 𝍸 𝍸 𝍸 𝍸 𝍸
Total Red Flowers:	Total White Flowers:	Total Pink Flowers:

Table 4.2: F_1 Cross Results for Sweet Pea Flower Color:

Red Flowers	White Flowers	Pink Flowers
Total Red Flowers:	Total White Flowers:	Total Pink Flowers:
Ratios: (red:pink:white)		

Table 4.3: Chi-Square Calculations

Phenotype	Observed (o)	Expected (e)	o - e	$(o - e)^2$
Total				

CONCLUSIONS

Experiment 1

1. What pea color is dominant? Recessive? How did you arrive at this conclusion?

2. Using "C" and "c", what are the genotypes of the P_1 and F_1 plants?

3. What ratio of yellow and green peas would you expect in the F_2 generation?

4. By doing a Chi Square analysis on the F_2 data, can you confirm that these plants are a Mendelian F_2 Generation? Show your work.

Experiment 2

1. What flower position is dominant? Recessive? How did you arrive at this conclusion?

2. Using "P" and "p", what are the genotypes of the P_1 and F_1 plants?

3. What ratio of axial and terminal flowers would you expect in the F_2 generation?

4. By doing a Chi Square analysis on the F_2 data, can you confirm that these plants are a Mendelian F_2 Generation? Show your work.

5. In the natural setting, the trait that is most common is called the wild-type and the rarer type is the mutant. With respect to Mendel's peas, green peas and axial flower position are the wild-types. By examining your data from Experiments 1 and 2:
    a. Which allele is dominant (expressed)?
    b. Which allele is the wild type?
    c. A mutation results in a phenotype that is different from the wild type. Dominant means that if you have inherited that allele you have that phenotype. Do mutations always cause an dominant allele to become a recessive allele? Support your reason with data.

Experiment 3

1. Do the genes for pea color and flower position exhibit Independent Assortment? How do you know this?

2. By doing a Chi Square analysis on the F_2 data, can you confirm that these plants are a Mendelian F_2 Generation showing Independent Assortment? Show your work

EXPERIMENT 16:
VIRTUAL GENETICS LAB

Objectives

1. Define the following:
    a. autosomal dominant trait
    b. autosomal recessive trait
    c. incomplete dominance
    d. X-linked recessive trait
2. When presented with data from controlled breedings, be able to determine the inheritance pattern.

Introduction

You will be presented with an imaginary organism that has various characteristics. It will be your job to determine which of the following patterns will explain the mode of inheritance for that characteristic.

- Autosomal Dominant/Recessive: Since the gene is on an autosome, there is an equal chance of males and females having the trait. When the trait is recessive, the individual must be homozygous for the allele and the parents may have the alternative phenotype. When the trait is dominant, for the offspring to have the trait at least one of the parents must also have the phenotype.

- Autosomal Incomplete Dominance: The parents have contracting phenotypes and the offspring have an intermediate phenotype.

- Sex Linked Recessive: Females are XX and males and XY. The gene is found on the X chromosome so the phenotype is more common in the male and females may be carriers.

- Sex Linked Incomplete dominance: One phenotype will only be found in females.

Virtual Genetics Lab (VGL) is a computer simulation of the genetics of an imaginary insect. The computer randomly picks a trait with two forms. It then randomly chooses which form will be dominant and which will be recessive. That way, each time you start the program, you and all the groups will receive a different problem. Finally, the simulation creates a population of insects (Field Population) with random genotypes. The first problem you will work will show the mode of inheritance and the genotypes. After you have become comfortable with the simulation, you will work two additional problems without help from the program.

cozyx
in - keep
ex - remove

not high
int

The program will generate the following patterns:

Autosomal alleles found in males and females	**Sex (X)-linked** alleles more likely in 1 sex
Complete Dominance 2 phenotypes	Complete Dominance 2 phenotypes recessive more likely in males
Incomplete Dominance 3 phenotypes	Incomplete Dominance 3 phenotypes found only in females

Materials and Methods

Adapted from http://intro.bio.umb.edu/VGL/index.htm

Materials

Computer with Virtual Genetics Lab (VGL)

Procedure

1. From the Desktop, open the VGL program. (If asked, click on the Run Program.) It is easy to fill the screen with cages of creatures and get totally confused, so work slowly, deliberately and keep careful notes about the experiments you do and the contents of each cage. If you get too confused, quit the problem by going to File, Close Work, Do Not Save and then start with a new problem.
2. To start a problem, Click on File, New Problem, Class Set 1. Each time you start a new problem, the computer will choose a new set of traits and inheritance pattern. A character with the same name may have a different inheritance pattern in a new problem.
3. Select "Show model and all genotypes" and click ok. The problems in Class Set 1 will give you the practice that you need to solve the problems in the other class sets. Class Set 1 are the only problems where you know the model and genotypes as you proceed through the problems.
4. The cage will appear (Figure 16.1) holding the "Field Population".

Figure 16.1: Initial Breeding Cage

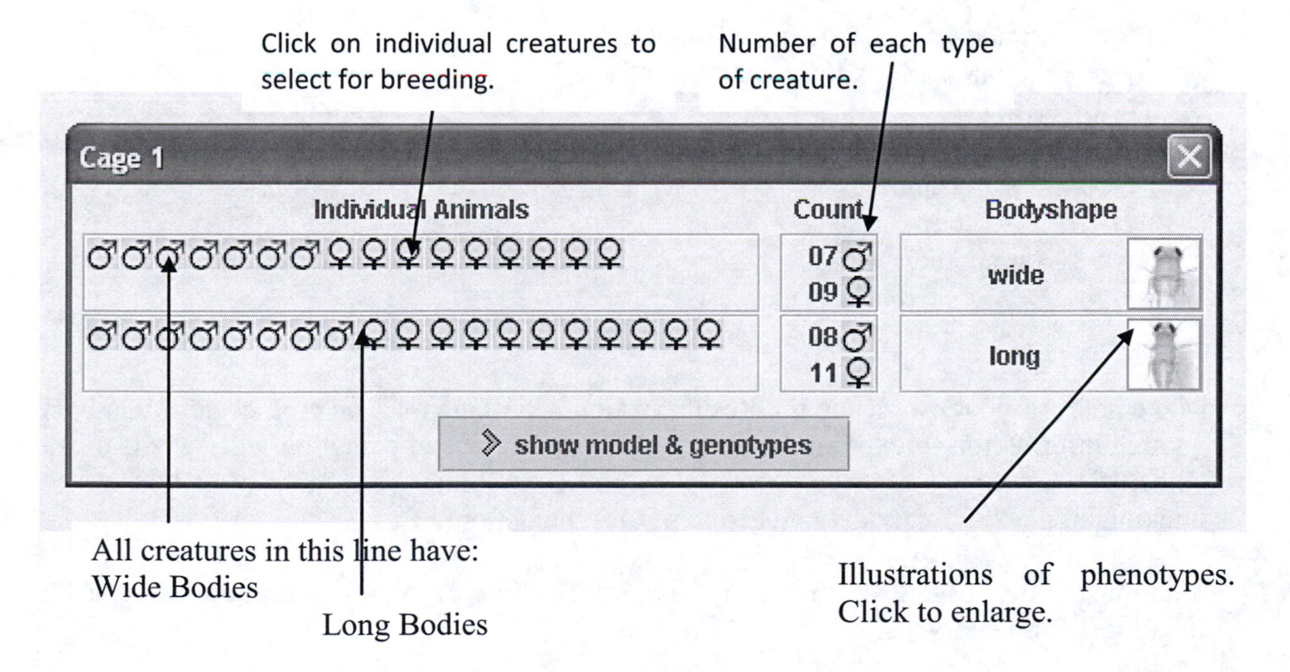

Your cage will be probably be different from the illustration. All problems in Class Set 1 are Autosomal Dominant/Recessive Patterns. By designing crosses and analyzing your results, you will be able to determine the dominant and recessive traits.

5. To begin an experiment, select the parents (a male and female) by clicking on the organism. (As you continue with the experiment, the parents may come from different cages and an individual fly may be bred more than once.) Once you have selected the parents, click on "Cross" and a new cage (In this case, Cage 2) will appear. You may need to drag the cages apart. **HINT:** Make sure you breed parents with the same phenotypes—breed like to like with the expectation that different phenotypes will result in the offspring.

Figure 16.2: Results of First Cross

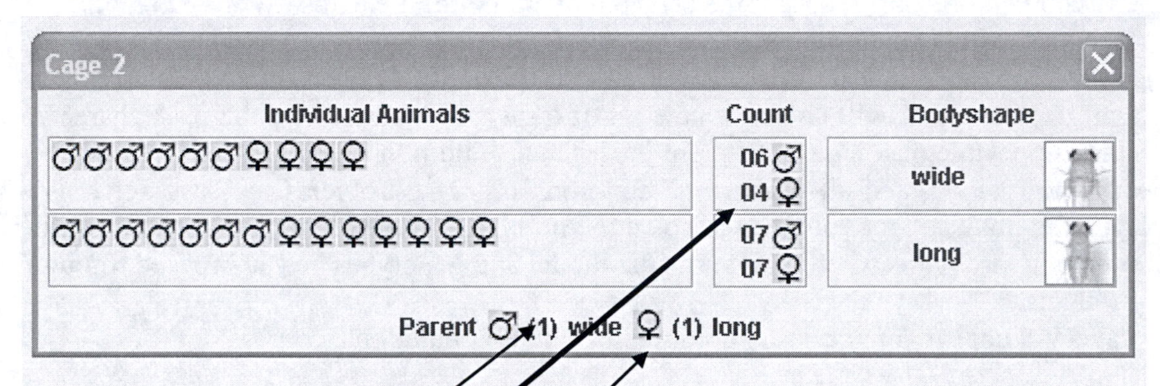

The cage is giving you the following information:

- The parents
  - A male with a wide body from Cage 1
  - A female with a long body from Cage 1
- The offspring
  - Wide bodies
    - 4 females
    - 6 males
  - Long bodies
    - 7 females
    - 7 males

6. Make notes of your crosses in the Result Section. Continue making crosses until you have enough information to explain the inheritance pattern (The explanation will be part of the Conclusion Section.) After you have solved the 1st problem, confirm your conclusion by clicking on the "Show model and genotypes" at the bottom of Cage 1.

7. You must know solve 2 problems in Class Set 2. These problems will not allow you see the model and genotypes.

Results

Problem 1 (from Class Set 1)

In the offspring phenotype columns make sure you indicate your phenotypes. If you need more crosses, extend the chart.

	PARENTS			OFFSPRING		
Cross #					Phenotypes	
		Cage #	Phenotype			
1	Male			Male		
	Female			Female		
2	Male			Male		
	Female			Female		
3	Male			Male		
	Female			Female		
4	Male			Male		
	Female			Female		
5	Male			Male		
	Female			Female		
6	Male			Male		
	Female			Female		
7	Male			Male		
	Female			Female		

Problem 2 (from Class Set 2)

In the offspring phenotype columns make sure you indicate your phenotypes. If you need more crosses, extend the chart.

	PARENTS			OFFSPRING			
Cross #					Phenotypes		
		Cage #	Phenotype				
1	Male			Male			
	Female			Female			
2	Male			Male			
	Female			Female			
3	Male			Male			
	Female			Female			
4	Male			Male			
	Female			Female			
5	Male			Male			
	Female			Female			
6	Male			Male			
	Female			Female			
7	Male			Male			
	Female			Female			

Problem 3 (from Class Set 2)

In the offspring phenotype columns make sure you indicate your phenotypes. If you need more crosses, extend the chart.

	PARENTS			**OFFSPRING**			
Cross #					Phenotypes		
		Cage #	Phenotype				
1	Male			Male			
	Female			Female			
2	Male			Male			
	Female			Female			
3	Male			Male			
	Female			Female			
4	Male			Male			
	Female			Female			
5	Male			Male			
	Female			Female			
6	Male			Male			
	Female			Female			
7	Male			Male			
	Female			Female			

Conclusions

1. Summarize the results for each problem in the following table.

Characteristic	Mode of Inheritance*	Dominant trait	Recessive trait	Heterozygote phenotype
1.				
2.				
3.				

* Chromosome: autosome or sex-linked
  Allele: simple dominance or incomplete dominance

2. How did you determine the mode of inheritance? For each experiment write a short paragraph describing the cross(s) that allowed you to determine the mode of inheritance. Make sure to you refer to the phenotypes of the parents and offspring to support your conclusions. Write the genotypes for these parents. A series of Punnet squares may be used to support your conclusions.

EXPERIMENT 17: HUMAN KARYOTYPES

Objectives

1. Define the following: karyotype, autosome, sex chromosomes, non-disjunction, aneuploid.
2. For humans, state the total number of chromosomes, number of autosomes and sex chromosomes in a somatic cell.
3. Given an individual's karyotype and a key be able to determine the phenotype of the individual.
4. Explain how non-disjunction leads to an aneuploid condition.

Introduction

Somatic cells contain the diploid number of chromosomes (are in pairs) and are produced by mitosis. In humans, this means that each cell has a total of 23 pairs. These chromosomes are classified as autosomes (22 pairs) and one set of sex chromosomes. A male has an X and a Y while a female has 2 X's.

Gametes contain the haploid number of chromosomes (one of each pair) and are produced by meiosis. At times, errors occur where both of the chromosome pairs end up in one cell (therefore another cell will receive none of that pair). Such errors are caused by non-disjunction. If a gamete with an abnormal number of chromosomes is involved in fertilization, the resulting zygote will have an abnormal chromosome number (called an aneuploid). When the zygote is missing a chromosome the condition is referred to as a monosomy condition and when there is an additional chromosome, it is a trisomy. The resulting individual will be minimally to severely affected, depending on which chromosome is affected. (Table 17.1)

Table 17.1: Examples of Chromosomal Abnormalities

ABNORMALITY	EFFECT
Trisomy 18	Edwards Syndrome: usually fatal within 3 mo. due to multiple congenital defects
Trisomy 13	Patau Syndrome: numerous physical abnormalities, especially heart defects and incomplete brains; generally death in infancy
Trisomy 21	Downs' Syndrome: mental retardation, short & curved 5th finger, marked creases in palm, characteristic facial appearance, heart defects
XXY	Klinefelter syndrome: "male" with underdeveloped testes, enlarged breasts, usually sterile
XYY	XYY male: tall, persistent acne, fertile
XO	Turners: "female" with underdeveloped ovaries, doesn't go through puberty
XXX	Metafemale: "female" with limited fertility

Cytogeneticists are able to analyze and map an individual's chromosomes. One method is the process of culturing dividing cells (white blood cells or fetal cells harvested from amniotic fluid or chorionic villi tissue) and arresting them in mitotic metaphase. At this particular stage the DNA (chromatin material) has doubled, the strands are connected at the centromere, are highly condensed and now is called a chromosome. The cells are stained and observed under a microscope. Pictures are taken of these cells and the chromosomes are paired and arranged by size, centromere location and banding patterns generated by the stain. (Table 17.2)

Table 17.2: Chromosome Groups

CHROMOSOME GROUP	CHROMOSOME NUMBERS	DESCRIPTION
A	1, 2, 3	Very long arms with median centromeres
B	4, 5	Long with submedian centromeres
C	6, 7, 8, 9, 10, 11, 12	Median with submedian centromeres
D	13, 14, 15	Median with terminal centromeres
E	16, 17, 18	Short and median/submedian centromeres
F	19, 20	Very short with submedian centromeres
G	21, 22	Very short with terminal centromeres
Sex	XX or XY	

The resulting "picture" of the chromosomes is called a karyotype. (Figures 17.1a and 17.1b)

Figure 17.1a: Karyotype of Normal Female

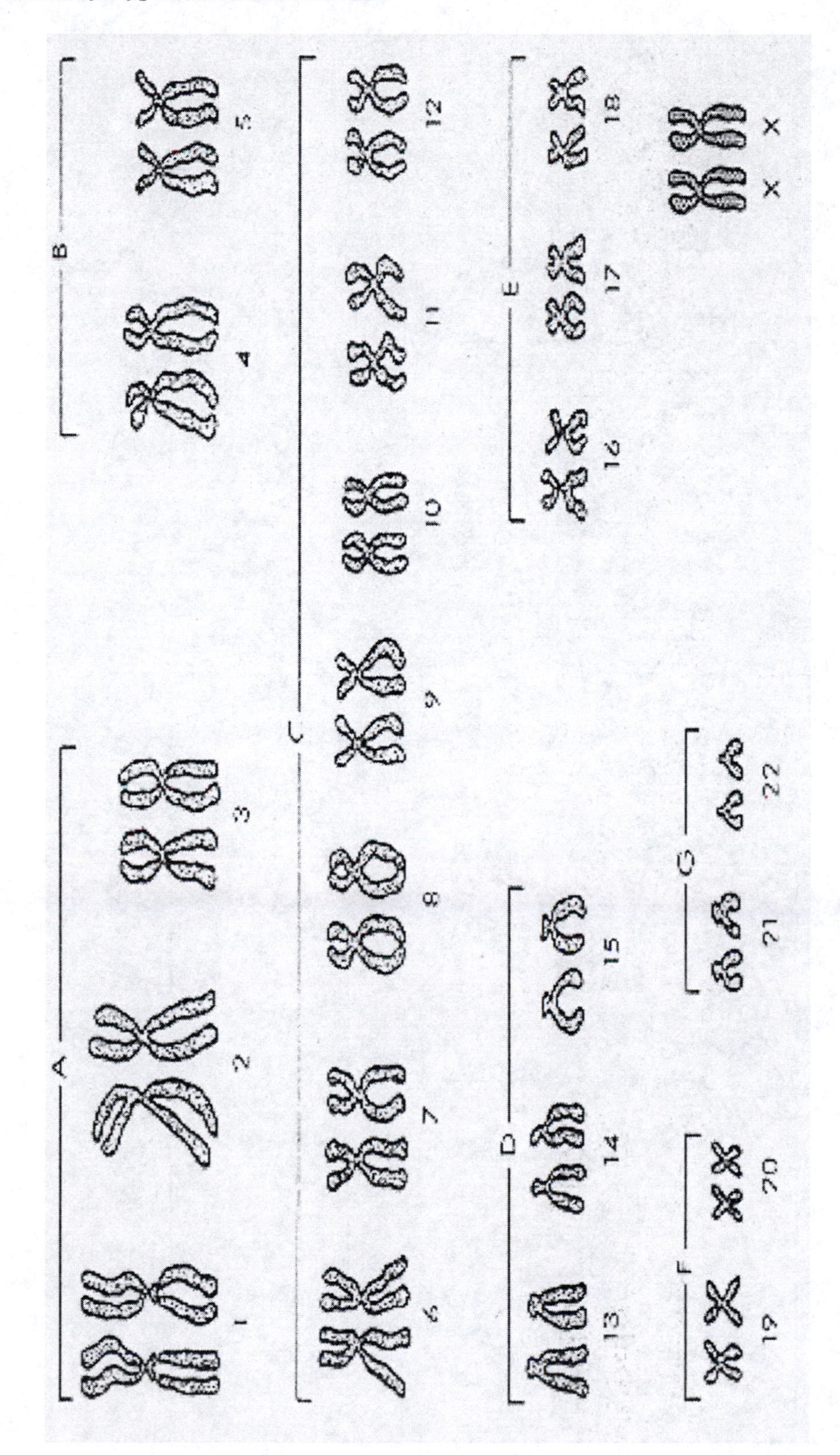

Figure 17.1b: Karyotype of Normal Male

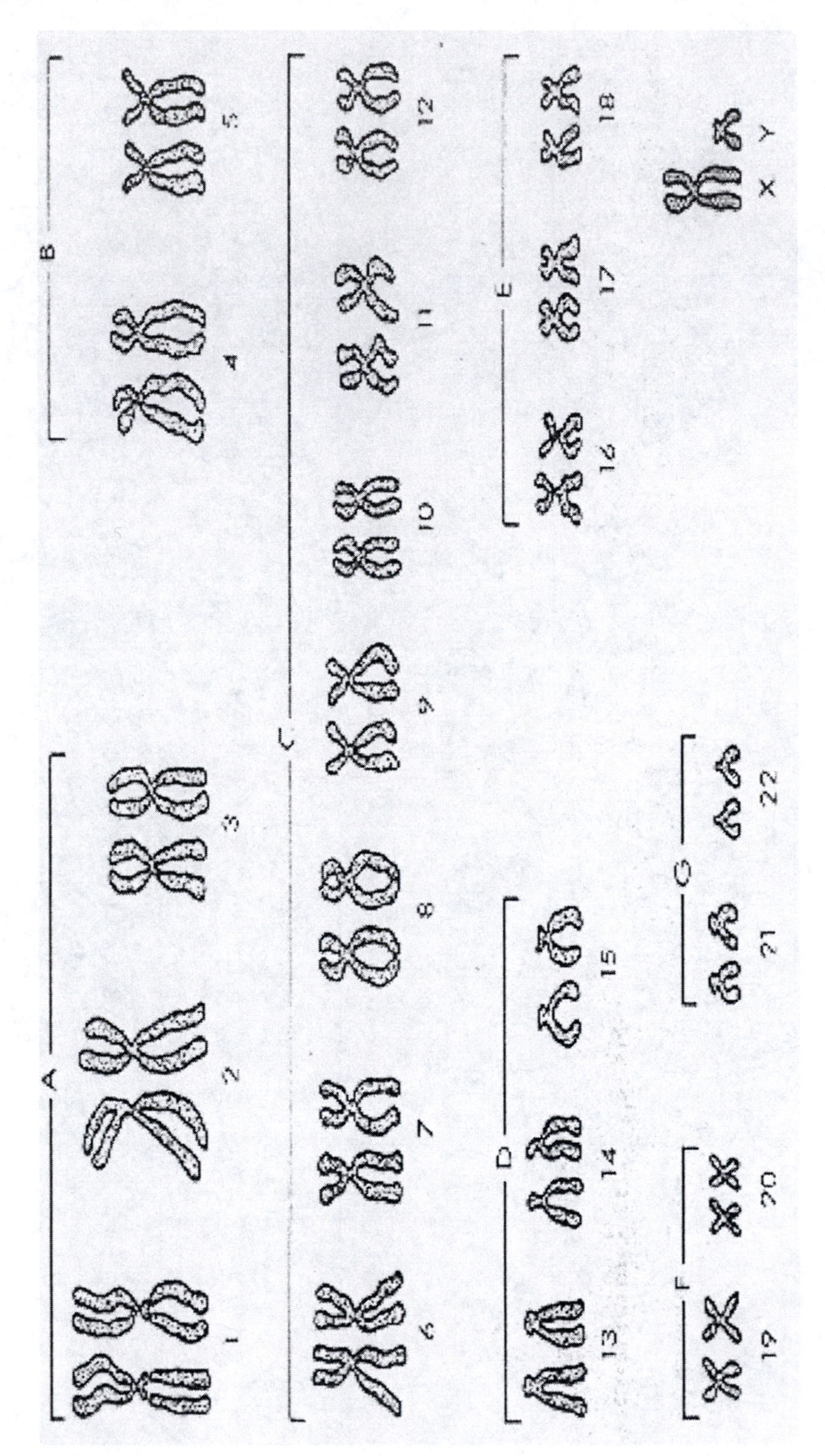

Materials and Methods

Materials

- scissors
- tape

Procedure

1. Each individual in the group is responsible for preparing one of the karyotypes.
2. Carefully cut out the chromosomes from the worksheet.
3. Using the pictures in Figures 17.1a-b, arrange the chromosomes in the proper location on the worksheet.
4. Once you have completed the karyotype determine the condition of the person with this type of cell.
5. Obtain the "diagnosis" of the other karyotypes in your group.

Individual 1

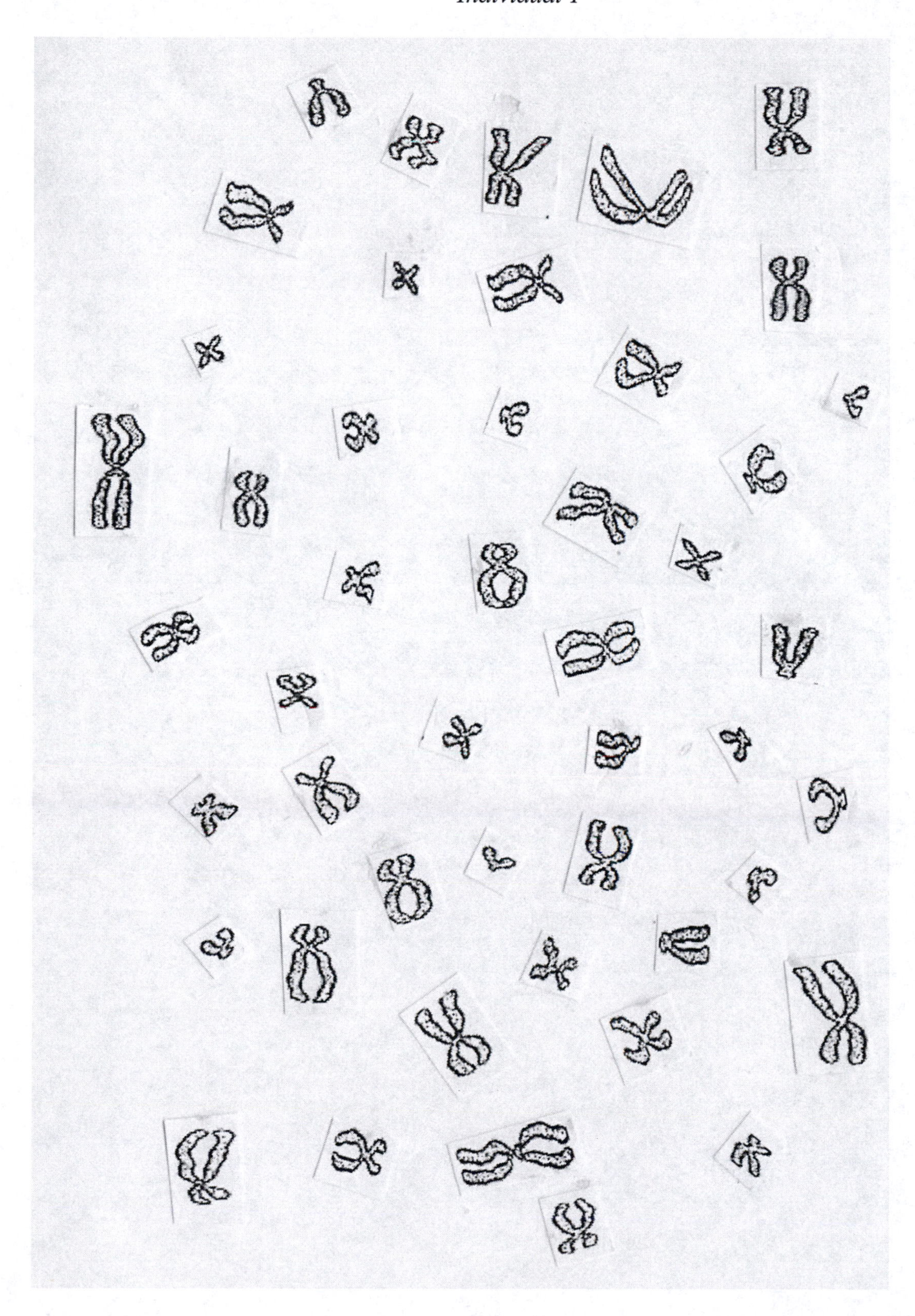

Individual 2

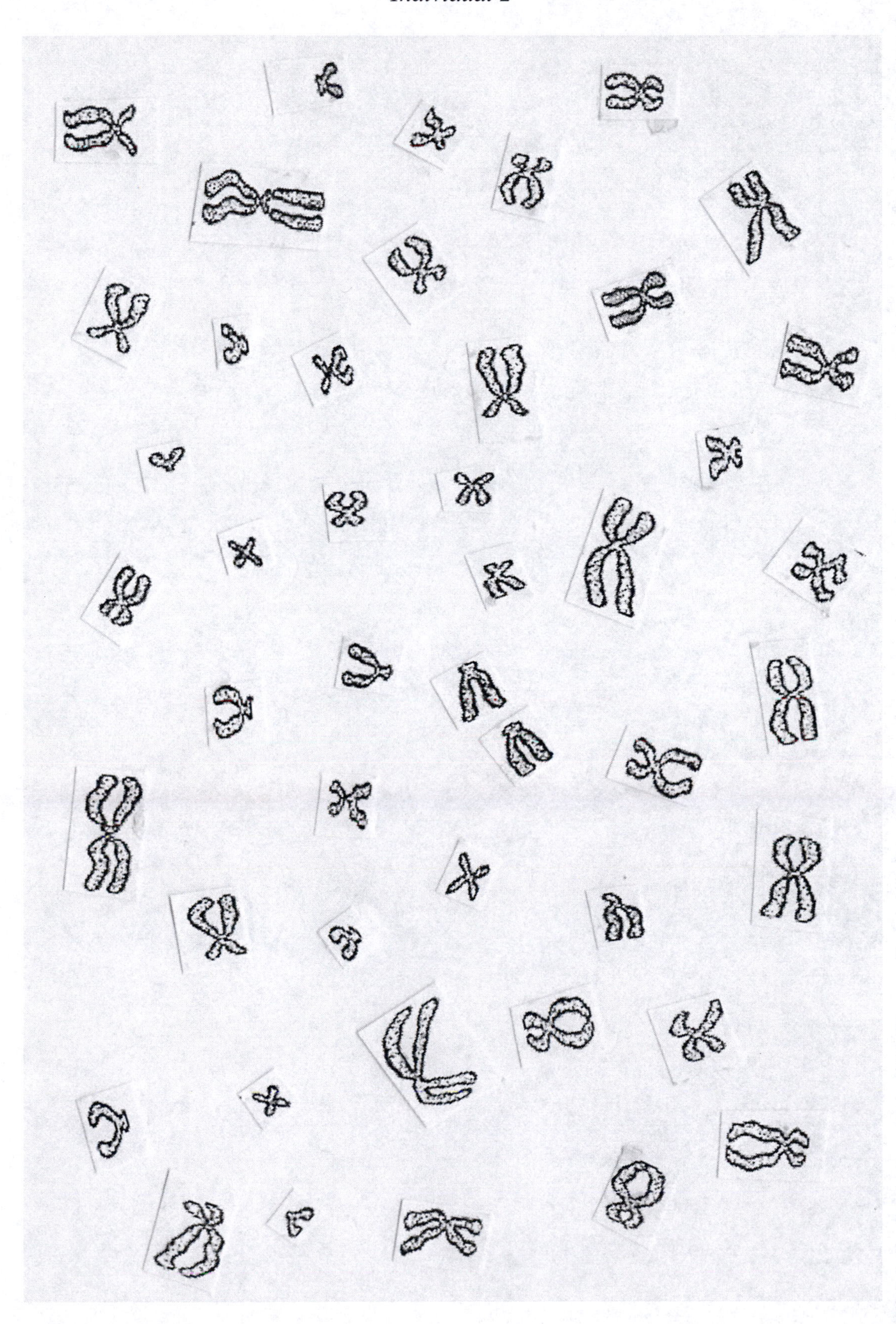

Individual 3

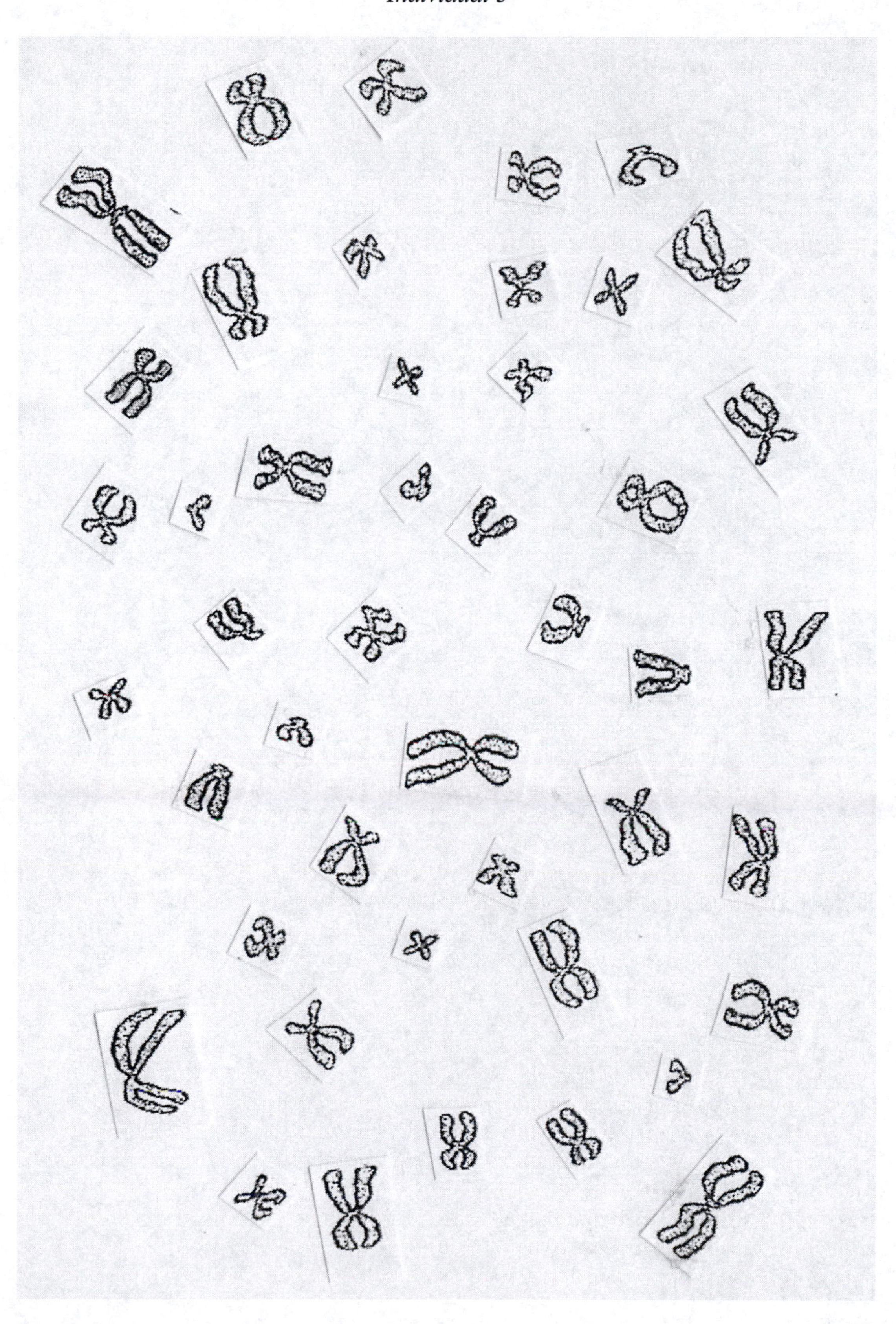

Individual 4

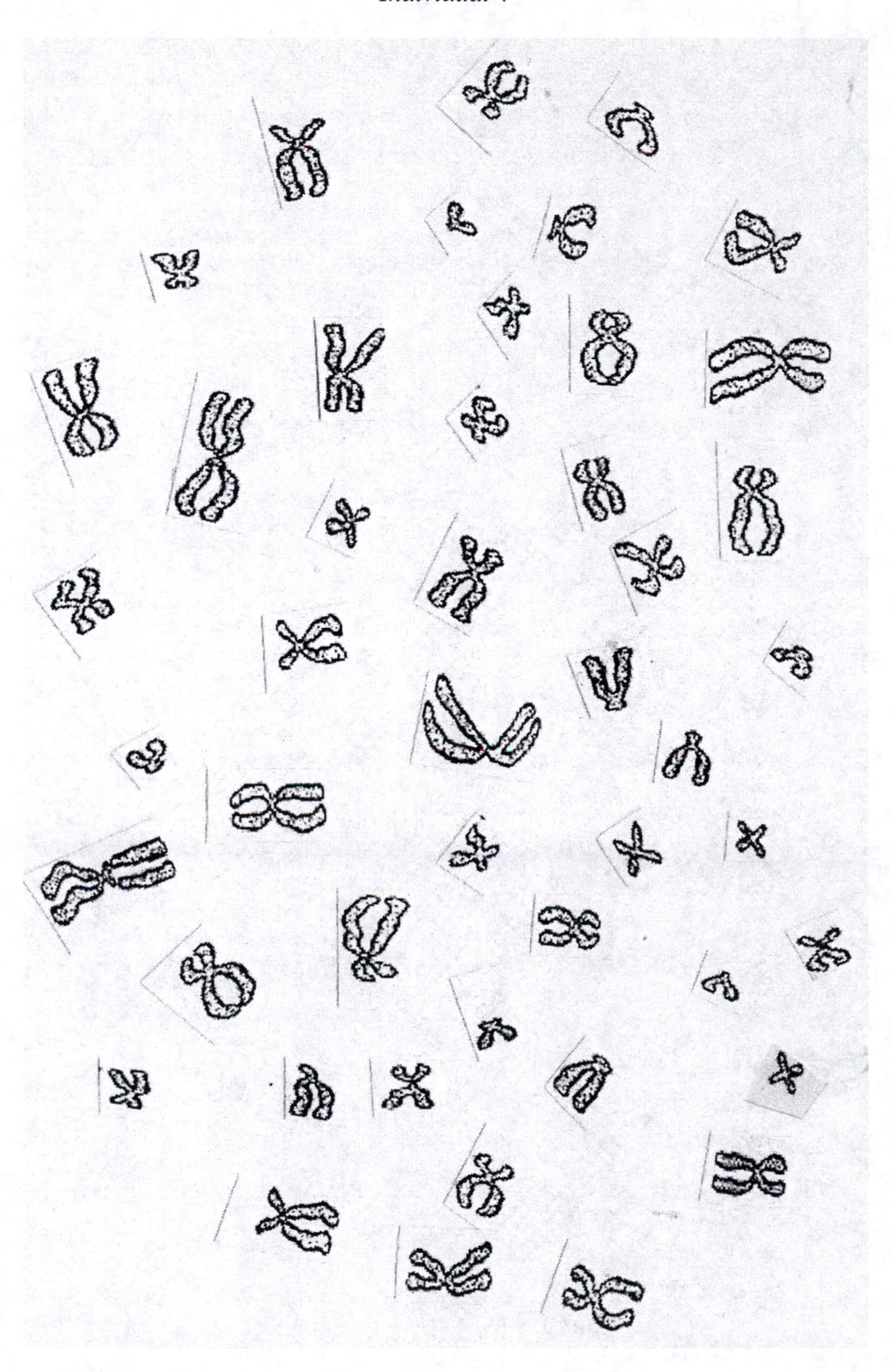

Results

1	2	3
4	**5**	**6**
7	**8**	**9**
10	**11**	**12**
13	**14**	**15**
16	**17**	**18**

19	20	21
22	**Sex**	

Individual #: _________

Sex of subject: __________

Number of chromosomes: __________

Condition of subject: ____

Summary Chart

Individual	Sex	# Chromosomes	Condition
1			
2			
3			
4			

Conclusions

1. Why must mitotic cells be used to prepare a karyotype?

2. What is the difference between a monosomy and trisomy individual? Explain, at the cellular level, how these conditions are created.

3. Human autosomal aneuploids generally die *in utero* or in early childhood. However, aneuploid conditions involving the sex chromosomes can have a normal life expectancy. Why?

APPENDIX A
SCIENCE SAFETY PROCEDURES

To participate in BIOL 110 labs, you must view the Safety Power Point presentation and pass the safety quiz. Below is a summary of the major points contained in the presentation.

Every effort has been made by the faculty and staff at CCBC to insure that your laboratory experience here is a safe one. You must be aware that much of the responsibility for making the lab a safe place to work and learn rests on your shoulders. There are potential hazards involved with laboratory work that can lead to accidents and injuries; therefore, the following safety rules and regulations are of the utmost importance and must be adhered to at all times.

- In case of an emergency, call 911 then Public Safety (443-840-1111 or the button on the wall pixie)

- DO NOT eat, drink, smoke, or chew in the lab. Do not put anything in or on your mouth; this includes pens, pencils, lip gloss, lip stick and pipettes.

- If you do not know what something is, treat it as if it were dangerous until you know otherwise.

- Clothing requirements
  - Wear closed toed flat shoes that cover the tops of your feet. The shoes should provide good traction.
  - One of the following:
    - Preferred: A buttoned lab coat that comes down to your knees.
    - Legs must be covered to the knees with no skin showing, no bare midriffs or no tank tops. You must wear at least a short sleeve shirt.

- All wet labs require the proper use of safety goggles. If you are handling hazardous materials please wear gloves.

- For most laboratory exercises, hazardous waste materials should be placed in special waste containers and not flushed down the drain or placed in a wastebasket. These disposal containers are generally located in the front of the room. When special disposal is needed, your instructor will point their location. If you are unsure of the proper means of disposing of materials, you should consult the instructor or lab technician before disposing of any solids, or flushing anything down the drain.

- Handle glass carefully. Never force or stress glass as it can break and cause serious cuts. Also, be sure to use proper protection when handling hot glass (hot pads etc).
  - Broken glassware, used pipettes, sharps (i.e. scalpel blades) and broken slides must be disposed of in the properly labeled receptacles. The instructor should be notified of any broke glassware or slides. Non-hazardous trash must not be thrown in here; place trash in the trash can. Be sure to inform the instructor.

- Keep your work area clear. To prevent damage to your personal effects store book bags and clothes elsewhere.

- Keep water and conducting materials away from electrical outlets.

- Do not run in lab. Horseplay is prohibited.

- Only registered BIOL 110 students that have passed the safety quiz are permitted in the lab.

- Please read the lab exercises carefully. Do not perform "unauthorized" experiments or try out a new idea unless the instructor approves it.

- Return all items such as slides, models, equipment, supplies, or microscopes to their proper location.

- Clean your lab bench before leaving and wash your hands thoroughly with soap and water, as appropriate.

- Push the stools under the benches when you leave.

APPENDIX B
GUIDE TO WRITING A FORMAL SCIENTIFIC PAPER

Why

It is assumed that written material will be read. The writer then must guide his writing to the reader most likely to read that material. Your lab manual has been written for students who have a fundamental knowledge of each particular lab subject; but lack a detailed understanding of the material. Your paper is also aimed to this audience—someone who understands basic science; but has very little knowledge of your topic. In other words, assume your reader knows very little if anything—explain everything.

The report will follow the steps of the Scientific Method. Start by reviewing your weekly lab reports: you will need the question/purpose of the experiment, the hypothesis, correctly collected data, graphs and the conclusion question(s) that explained your data. You will then add to this information. Read the sample lab report that follows these directions.

Sections

Title (placed on a Title page)

The title should indicate what the paper is about. The title needs to be very specific and is a simple declarative statement. Do not make the title a question or a cute saying. The title is the question/purpose of your experiment. For example, you are developing a new artificial sweetener, X, and are looking to see if it causes bladder cancer in laboratory mice. An appropriate title for the paper might be: "The Effect of Artificial Sweetener, X, on the Development of Bladder Cancer in Laboratory Mice". After reading the title, a researcher interested in artificial sweeteners or bladder cancer might want to read the article.

The title page also needs your name, your lab section and the date the report is due.

Abstract

This is a brief summary of the entire report. In one paragraph (5 to 7 sentences) state why you are doing the experiment, how you did it, what happened and why it occurred. This section is not easy to do and although it comes first in the paper, you generally compose it after the entire paper has been completed. Examine the samples that follow these directions.

Introduction

This section contains the statement of the specific question(s) being studied, hypothesis and the relevant scientific background information. (Examine the Introduction Hints that follow. Do not include information that is not necessary to understand the experiment.) Background information will be obtained through library research. Since the paper's audience is other students, make sure you define all biologic terms. Since your audience is a beginning biology student, make sure you define all scientific terms.

Materials and Methods

You will NOT need to include the procedure in your report. Instead, include the following statement in this section: "Procedure is found on pages _____ to _____ in the BIOL 110 Lab Manual."

Results

This needs to include what happened in the experiment. Numerical data is presented in tables and if also appropriate, graphs. Tables are consecutively numbered and titled. (example: Table 1: Title) Each table is typed on a separate page and is placed at the end of the Results Section. Graphs, drawing, etc. are called figures. They are also numbered, titled and placed on separate pages in the Results Section.

Since you are not including a methods section, tell how the data was collected and include any formulas that were used to analyze the data.

Discussion and Conclusion

In the Introduction, you asked a question. The answer to this question is found in the data you collected. It is in this section where you results are explained. Make sure you use data to explain your conclusions. Explain the reasoning you used to arrive at your conclusion.

- Did your experiment support or refute the hypothesis?
- How does the information contained in the introduction connect to your results?
- Are your conclusions supported with data? Use actual numbers and refer to graphs.
- Do your results agree with established biological facts? If not, what do you think went wrong with your experiment? Student error is not an acceptable explanation.

Relevance

In this section you will explain how this experiment relates to something in your everyday life. To do this you will need to find an article or web site from a non-science, non-academic source. Summarize the article and explain how your conclusions and knowledge obtained from the experiment can relate to the article.

Literature Cited

You must have at least four references, one of which must be this manual and no more than 1 web site may be used. If you find an article on the web and the article is also available in print form, then the site does not count as a web site. List sources you used for information in your report and use proper APA formatting. (See below.)

Web Site Requirements

- You may not Wikipedia or any general type of encyclopedia; but scientific encyclopedias are permitted.
- Dictionaries are not permitted.
- Sites must be appropriate academic resources.
- You may not reference any classroom Power Point presentations or online lecture notes.
- You may use no more than 2 textbooks (Your lab manual and an encyclopedia are textbooks.).

How to Reference

While you must use your own words to explain your investigation, you still must reference the material that gave you the information. To reference the source in the body of the report, you will the parenthetical format. (See below.)

Direct quotes should be limited to 1 or 2 lines and used sparingly. It is your job to show that you understand the material that you are presenting, so you MUST use your own words. Remember that NONE of the information contained in the introduction (and possibly other sections) is your own original material—it came from someone else. Therefore, it must be properly referenced (even if paraphrased material); if not, this is also plagiarism—you stole someone else's ideas.

Plagiarism is to use material obtained from other sources (journals, textbooks, web sites) and represent it is as your own work. It is imperative that your paper be in your own words and that you indicate where you found the information that you are using in your paper. Please refer to the following web site (http://ori.hhs.gov/education/products/roig_st_johns/Introduction.html) for detailed information on how to avoid plagiarism.

Formatting rules:

- Margins must be between 1-1 ½ inches.
- Use Times Roman, Ariel or Courier type fonts in either 10 or 12 point size.
- Use double spacing.
- Each section is titled-example Abstract, Results and each section should start on its own page.
- Properly numbered and titled graphs and charts are to be placed into the Results section, not at the end of the report.
- The entire paper is to be written in 3rd person, past tense.
- Include a grading sheet with your report.

CCBC On-Line Library Databases

Getting Access

The on-line databases are available from on and off campus computers

- From the CCBC main web page (http://www.ccbcmd.edu) click on the Current Student Tab
  - Libraries
    - Science and Math/Biology
- Click on Find Magazines, Journals and Newspapers
- To log-in you will need your student ID number.

Useful Databases

ProQuest: Contains a full collection of journals, magazines and newspapers for information on a broad range of general reference subjects.

Health Course: Consumer Edition: Information on many health topics. Full text articles are included as well as abstracts and indexing for nearly 180 general health publications.

Science Resource Center: Scientific developments, topic overviews, experiments, biographies, pictures and illustrations are covered in articles from over 200 magazines and academic journals and links to quality web sites.

Science Direct: contains full-text articles, book series, references works and abstracts in science, technology and medicine

Helpful Hints

- Many of the data bases have help sections
- Have the librarians help

Revising, Editing, Proofreading: Who, What, Where and When

Developed by Ann Kiby, English Department

Step in the Writing Process	Who	What	Where	When
Composing	The **author**	Formal lab report	Outside the class	After the lab has been completed.
Revising (requires a text)	The **author** only. (If you need help, someone will ask what you want to say.	Looking at **content** to make sure you include all the ideas and fact you intend to share with the reader.		After a draft is written with all its necessary parts.
Editing	**Anyone with adequate skills** in writing who understands the purpose of the writing—perhaps a member of your intended audience	Making sure that your **paper** will allow your audience to know what you want them to know as easily as possible—that you've clearly said what you meant to say.		When all revision are complete
Proofreading	**Almost anyone**	**Mechanics**—punctuation, graphic interface		When all other steps are completed—this is the last step before submission

American Psychological Association Style

Reference List Basics

- Indent ½ inch before the 2nd and all subsequent lines in the citation.
- Arrange the completed reference list in 1 alphabetical list by 1st author's last name.
- Give only the 1st city of publication, use official 2-letter US Postal Service abbreviations for states
- Only the 1st word of a title or subtitle and proper nouns are capitalized in book titles, magazines, journals and newspaper titles and subtitles. All major words are capitalized in journal titles.
- Each reference must end with a period unless the reference ends with a DOI or URL.
- Each reference cited in your text must appear in the reference list, and each resource on the reference list must be cited in the text.
- Use PDF version of electronic resources for page numbers.
- Double space all lines within and between entries. Examples provided are single spaced to save space.
- Italicize titles of books and journals.
- Use n.d. (no date) when there is no publication date.
- Caution: citations provided by databases are not always correct. Verify for accuracy.

Parenthetical references usually contain author's last name and year. If the author's name appears in the text, it may be omitted from the parenthetical reference.

Parenthetical References

Instead of footnotes or endnotes, use parenthetical references which are shorter or clearer. Remember that every parenthetical reference must have a corresponding full listing in the bibliography at the end of your paper. Examples are from the 6th edition of the *Publication manual of the American Psychological Association.* Washington, DC: American Psychological Association.

Author's Name in Text	Walker (2000) compared reaction times.
Article or Single Volume Book	In a recent study of reaction times (Walker, 2000)
Work Listed by Title and Year (No author given or anonymous)	on free care ("Study Finds," 1982)
Group as Author (Use full name in citation)	(University of Pittsburgh, 1993)
Quotations (Always include page numbers)	(Cheek and Buss, 1981, p.332)

Bibliography

The following examples are taken from the fifth edition (2001) of the *Publication Manual of the APA,* pp. 215-281. It includes more examples than are shown here. Please refer to it for detailed explanations on the various elements of the references.

<u>Print Sources</u>

Book by a Single Author

Last names of author, initial(s) of author. (Date). *Title of book.* City, State of publication. Publisher

Book by (up to 7) Multiple Authors

Last name of 1st author, initials(s) of author & Last name, initial(s) of 2nd author. (Date). *Title of book.* City, State of publication. Publisher.

Editor, no author

Last name of editor, initial(s) of editor. (Ed.). (Date). *Title of book.* City, State of publication. Publisher

Chapter in an Edited Book

Last name of article chapter author, initial(s) of author. (Date). Chapter title. In Book editor initial(s) and last name (Ed.) *Title of book* (pp. page numbers of chapter). City, State of publication. Published.

Government Publication

Name of Government Agency. (Date). *Title of publication.* City, State of publication. Publisher.

Encyclopedia Article

Last name of article author, initials(s). (Date). Title of article. Encyclopedia editor initials(s) and last name (Ed.), *Encyclopedia title.* (Vol. volume number, pp. page numbers of article) City, State of publication. Publisher.

<u>Magazines, Journals and Newspapers</u>

Magazine

Last name of author, initial(s). (Date in year, month day format). Title of article. *Magazine title,* Page(s).

Journal

Last name of author, initial(s). (Date). Title of article. *Journal title, Volume number* (issue number). Page(s).

Newspaper

Last name of author, initials(s). (Date in year, month day formal). Title of article. *Newspaper Title,* p. (pp) Page(s).

Electronic Resources

Articles from Library Databases

Citation for print version of the source. Digital object identifier (DOI) is available. retrieved from name of database.

Source with Author

Last name of author, initial(s) of author/editor. (Date created). *Title of site.* DOI if available. Retrieved from URL.

Source with Corporate Author

Name of institution. (Date created). *Title of site.* Retrieved from URL.

Source with No Author

Title of site. (Date created). Retrieved from URL.

Sample Lab Report

The Effect of Different Sugars on the Fermentation Rate in Yeast

Abstract:

Alcoholic fermentation is an anaerobic process (no oxygen is used) which yeast can use to convert the stored energy in organic molecules such as glucose to adenosine triphosphate (ATP) and the waste products ethanol and CO_2. An experiment was conducted to determine if various sugars were equally fermented by yeast. By monitoring a change in pressure due to the accumulation of CO_2, it was determined that yeast could effectively ferment glucose and sucrose while fructose was poorly fermented and lactose was not able to be fermented.

Introduction:

All cells require ATP to do work. To make ATP, cells must convert the energy that is stored in the covalent bonds of other organic macromolecules into the bonds found in ATP. Generally, cells have two methods available to make ATP. The more efficient method is to use oxygen. This aerobic process can be summarized as:

$$\text{organic molecule} + O_2 \rightarrow 36\ \text{ATP} + CO_2 + H_2O.$$

When oxygen is not available, cells, such as yeast, will need to generate ATP by an anaerobic process called fermentation. (Dalton, 2012) This process can be summarized as:

$$\text{organic molecule} \rightarrow 2\ \text{ATP} + \text{ethanol} + CO_2.$$

Many different macromolecules can be used as a source of ATP but the cells must have the proper enzymes to break the bonds in these macromolecules. This experiment investigated whether different sugars would be equally fermented by yeast. Glucose is a monosaccharide that is considered the primary energy source for cells because it can readily be converted into ATP. Fructose is an isomer (same molecular formula but a different structural formula) of glucose (Solomon, 2011) and is naturally found in fruits, vegetables and honey. (*Fructose Information Center,* 2012) Sucrose, commonly called table sugar, is a disaccharide composed a glucose molecule bonded to a fructose molecule and is used by plants to transport glucose throughout the structure. The last sugar studied was lactose, a disaccharide composed of two glucose molecules, is commonly found in milk. (Solomon, 2011) According to D'Amore et al (1988), glucose and fructose should have equal fermentation rates; however, when both sugars are present in equal concentrations, glucose is used at a higher rate than fructose thus it was assumed that sucrose would be fermented at a high rate.

Materials and Methods

Procedure is found on pages 189 to 190 in the BIOL 110 Lab Manual.

Results

Baker's yeast, *Saccharomyces cerevisae*, solutions were placed into test tubes and mixed with 5% solutions of the four sugars and kept in a 37°C anaerobic environment. The test tubes were connected to a Vernier gas pressure sensor which was then connected to a PC running the Vernier Logger Pro program. The CO_2 that was produced during fermentation would increase the pressure (measured in kilopascal) in the sensor. The program took frequent pressure readings, graphed the results and calculated the slope which was the reaction rate.

Table 1: Yeast Fermentation Rates

Contents of Test Tube	Fermentation Rate (kPa/min)
water (control)	0.0
glucose	2.0
fructose	0.5
sucrose	1.7
lactose	0.0

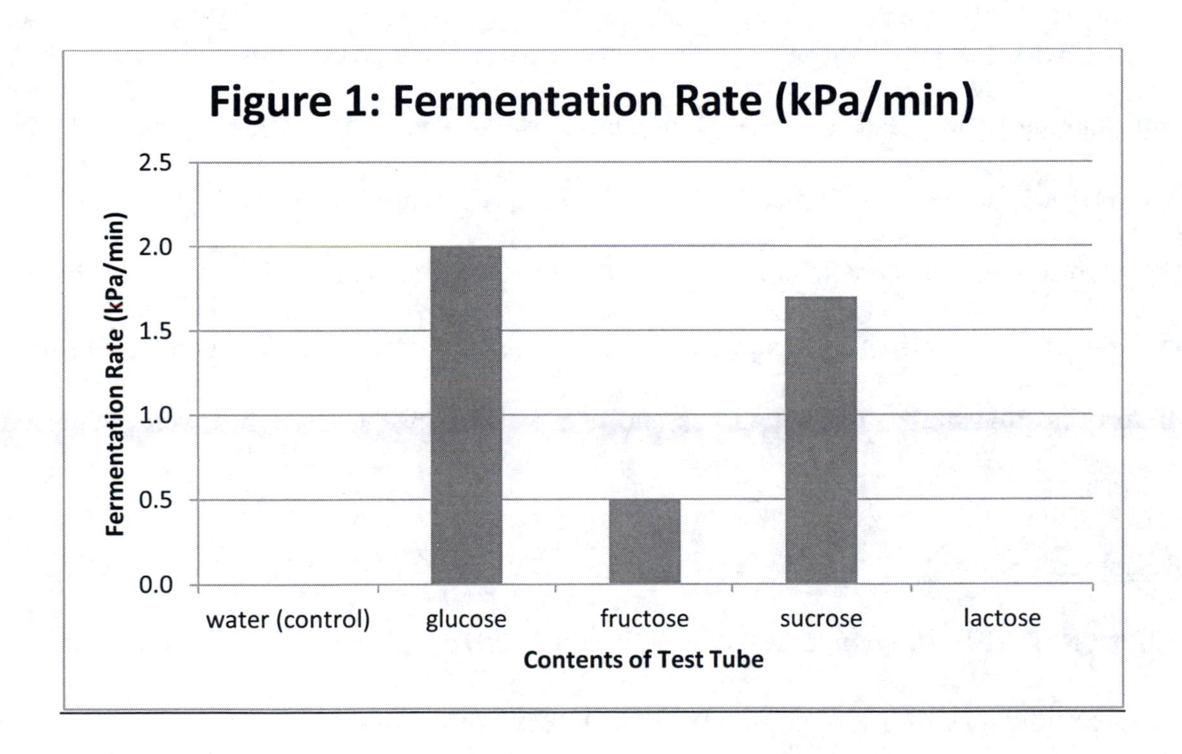

Conclusions

Of the four sugars tested, glucose was fermented at the highest rate (2.0 kPa/min) and lactose was not fermented at all. Glucose is the primary energy source for cells so it was expected that glucose would have the highest fermentation rate. Since yeast can be used to make alcoholic beverages such as wine and milk is not used in this process, it was expected that lactose would not be fermented. It was expected that fructose and glucose would have equal fermentation rates; however, fructose had a very slow rate (0.5 kPa/min). This was unexpected because wine is made from grapes which should have a high concentration of fructose (Solomon, 2011). The fermentation rate of sucrose (1.7 kPa/min) was almost as high as glucose. This rate confirms the work of D'Amore because sucrose is a disaccharide made of glucose and fructose. Once the bond holding the two monosaccharides together was broken, glucose would be readily available for fermentation.

Relevance

Many types of bread are produced by making dough with flour and yeast. The trick with making good bread is to get the dough to rise. For yeast to effectively make the bread rise, the dough needs to be placed into an environment that is between 80-115°F. The yeast will then ferment the contents of the dough and the CO_2 that is produced will make the dough increase in volume (rise) (Filippone, 2012). The reaction that was studied in this experiment is the process that makes the dough rise and it explains why the test tubes were incubated at 37°C (98°).

References

Dalton, K. (2012). *Biology I: Molecular & Cells, Lecture Guide & Laboratory Manual*. Virginia Beach, VA: Academx Publishing Services, Inc.

D'Amore, R. , I., & Stewart, G. G. (1989). Sugar utilization by yeast during fermentation. *Journal of Industrial Microbiology*, 4, 315-324.

Filippone, P. T. (2012). Leavening, what is yeast?. *About.com Home Cooking*. Retrieved from http://homecooking.about.com/od/specificfood/a/yeast.htm

Fructose Information Site. (2012) Retrieved from http://www.fructose.org

Solomon, E. P., Berg, L. R. & Martin, D. W. (2011). *Biology*. Belmont, CA. Brooks/Cole.

On Writing an Abstract

from http://research.berkeley.edu/ucday/abstract.html

An abstract is a short summary of your completed experiment. If done well, it makes the reader want to learn more about your research. These are the basic components of an abstract:

- Motivation/problem statement: What is the purpose of the experiment? What practical, scientific, or theoretical gap is your research filling? What is your hypothesis? What is the rationale of your hypothesis?
- Methods/procedure/approach: What did you actually do to get your results? (e.g. how did you collect data?, analyzed X samples, what did you change?)
- Results/findings: As a result of completing the above procedure, what did you learn? When possible use data to report your results.
- Conclusion/implications: What are the larger implications of your findings, especially for the problem/gap identified in step 1? Do the data support your hypothesis or must you reject it?

<u>Sample Abstracts:</u>

Title: Comparison of photosynthesis, transpiration, and water use efficiency in two desert shrubs.
Authors: Wei, Ruyi; Pan, Xiaoling

Abstract

Photosynthesis and transpiration are the most important physiological activities for plants. They are closely related not only to photosynthesis active radiation (PAR), air temperature (Ta), relative humility (RH) and ambient CO_2 concentration, but also to plant stomatal conductance and intercellular CO_2 concentration. Their relationship can be analyzed through statistic methods to indicate plant eco-physiological ability bearing to environment. By means of measuring the photosynthesis of *Haloxylon ammodendron* and *Reaumuria soongorica* in the Sangonghe River valley we have dealt with similarities and differences of two species. Results show that net photosynthesis rate of *Reaumuria soongorica* is slight higher than *Haloxylon ammodendron*, but water use efficiency of *Haloxylon ammodendron* is evidently higher than *Reaumuria soongorica*. It suggests that *Haloxylon ammodendron* has strong bearing ability to drought in natural habitat.

Title: Determination of Population Growth Rate and Carrying Capacity of Duckweeds (*Lemna minor*).
Author: CCBC student formal lab report

Abstract

Population growth and carrying capacity were determined for two cultures of *Lemna minor* (duckweed) over an approximate one month period. Culture 1 had an initial population (N_0) of 2 and culture 2 had an initial population of 15 duckweeds. The cultures were placed in separate flasks and maintained at constant temperature (approximately 23°C) and light availability. It was expected that culture 1 would initially have the faster intrinsic growth rate (*r*), and then slow to a similar rate of culture 2. The data collected periodically over the month period were somewhat contrasting to a typical growth curve, without clear log and stationary phases for either culture. Culture 1 exhibited a higher intrinsic growth rate of 1.5 individuals/[individual/month] and overall population growth rate (*dN/dt*) of 2 individuals/month than culture 2 (0.1 individuals/[individual/month] and 0.7 individuals/month, respectively). The higher rate for culture 1 are likely attributable to the Allee effect. The carrying capacity (*K*) was determined to be 6 duckweeds in culture 1 and 29 duckweeds in culture 2.

Title: The Effect of Light and Wind on the Transpiration Rate in Garlic Mustard (*Alliaria petiolata*).
Author: CCBC student formal lab report

Abstract

Transpiration is a process by which plants lose water through stomata of leaves to various environmental conditions. The water moves up through the xylem due to the special properties of water. A plants transpiration rate is closely related to the opening and closing of stomata which depends upon the environment. This experiment was performed in order to show how light and wind affect the transpiration rate in a dicot plant garlic mustard (*Alliaria petiolata*). It is expected that increased light will increase the transpiration rate because the increase in photosynthesis will require more water. It is expected that wind will also increase the transpiration rate because of wind lowers the water potential in the atmosphere. The results obtained from the experiment support the hypotheses. Light increased the transpiration rate to 2.1×10^{-5} kPa/ min/g and wind increased it to 3.9×10^{-5} kPa/ min/g.

Helpful Hints for Introductions

Osmosis in an Egg

- Define Brownian Motion and how it explains passive transport.
- Define osmosis and explain its importance to cells.
- Define solution, hypertonic, hypotonic and isotonic solutions.
- State the purpose of the experiment.
- What will happen to an egg when placed into each of the different "tonic" solutions.
- How will you measure the process?
- How will you use osmosis to determine the sugar content of the egg?

Enzymes

- What is the role of an enzyme and why are they important in cells?
- What is the chemical composition of enzymes and why is its shape important to its function?
- Explain denature.
- Explain the underlying principle behind all reactions (effective random collisions).
- What is the exact reaction and enzyme that was studied?
- What should happen to the reaction rate as enzyme concentration, temperature and pH changed? Why will the rate change?
- How will you measure the speed of the reaction? Discuss the role of guaicol, absorbance and the slope of the lines.

Photosynthesis

- State the general reaction.
- Completely explain the structure of the C3 leaf. (A diagram would be helpful.)
- Explain the solubility of the gases that are involved in the reaction.
- How will the structure of the leaf and the solubility of the gases be used to study the reaction?
- How do you get the segments to sink?
- How will the speed of the reaction be measured.
- What was your variable and the hypothesis?

Cellular Respiration

- State the general reaction.
- Explain the two Laws of Thermodynamics.
- Explain what the mouse needs to do to keep warm. Relate this to the molecular equation and the Laws of Thermodynamics.
- How will you get rid of the carbon dioxide?
- How will you measure the reaction rate?
- What where the two variables that were studies? State the hypothesis for each.

BIOL 110
Formal Lab Report
Grading Sheet

NAME: ____

Experiment: ____

Title and Formatting Instructions (3 points) ______
- Is the title appropriate to the experiment? ✓
- All sections labeled ✓

Abstract (6 points) ______
- Brief description of purpose, how, what and why happened ✓

Introduction (11 points) ______
- Appropriate background material—only what is necessary ✓
  - Do not include information not needed to understand experiment
- Purpose
- Hypothesis (expected results)
- How will you measure your results?

Results (8 points) ______
- Figures: Are they numbered and properly titled?
- Graphs: Are they numbered and properly titled? ✓
- Is there a brief description of how the data was collected? ✓

Discussion (12 points) ______
- Are the results explained?
- How do the results compare with your hypothesis?
- If it agrees, re-explain why it happened?
- If they do not agree, give possible explanation(s) for why not.

Relevance (7 points)
- Is the article summarized? ✓
- Did you explain why this experiment helps you understand an everyday occurrence? ✓
- Is a popular source used? ✓ ______

Bibliography (3 pts)
- At least 4 appropriate references ______
- Correct format. ✓
- Proper citing format within paper. ✓
- Is every source correctly sited? ✓

TOTAL ______/50

Lost Points
Excessive spelling and grammar mistakes.
Are the formatting directions followed.
Is the paper written in 3rd person, past tense.

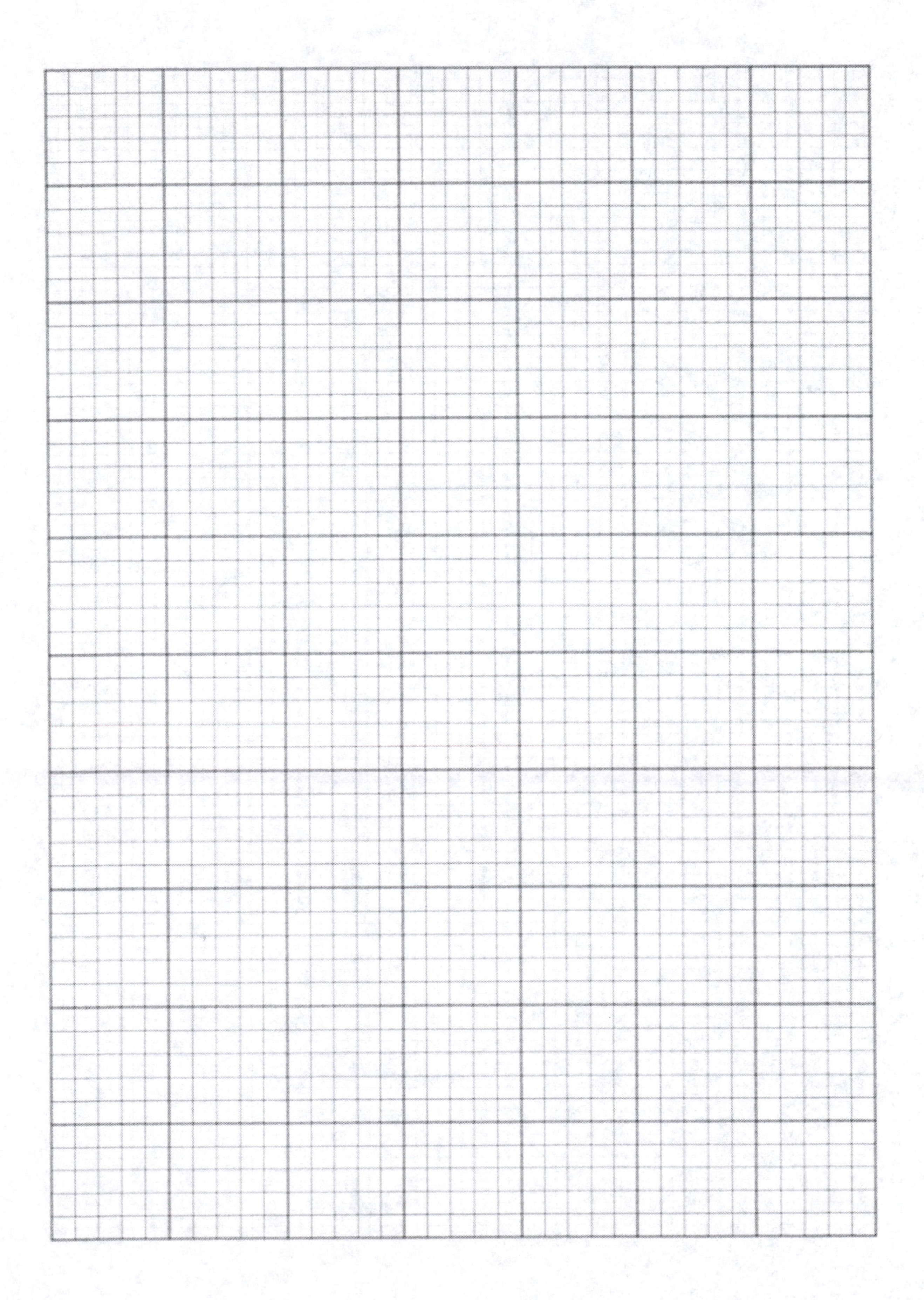

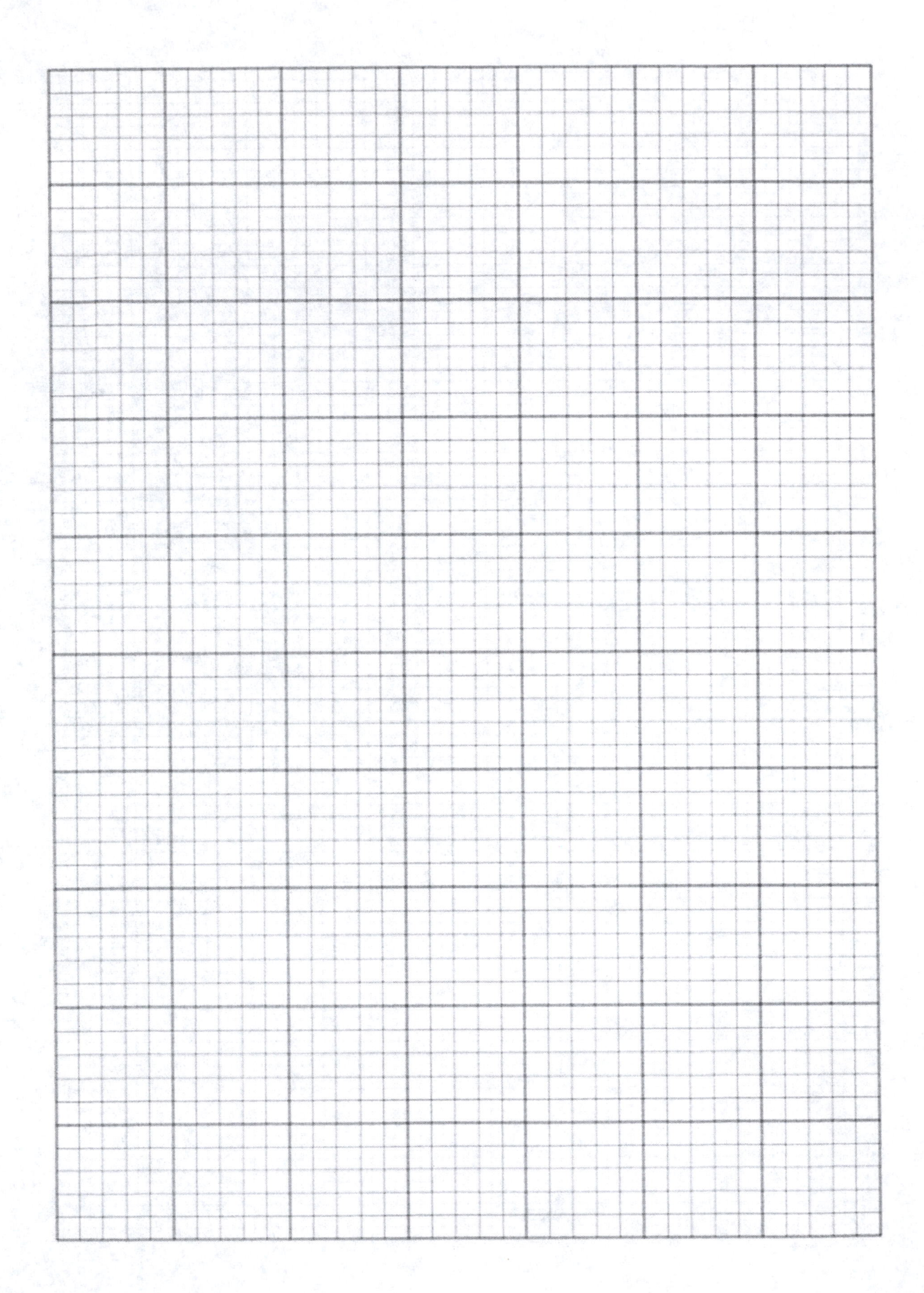

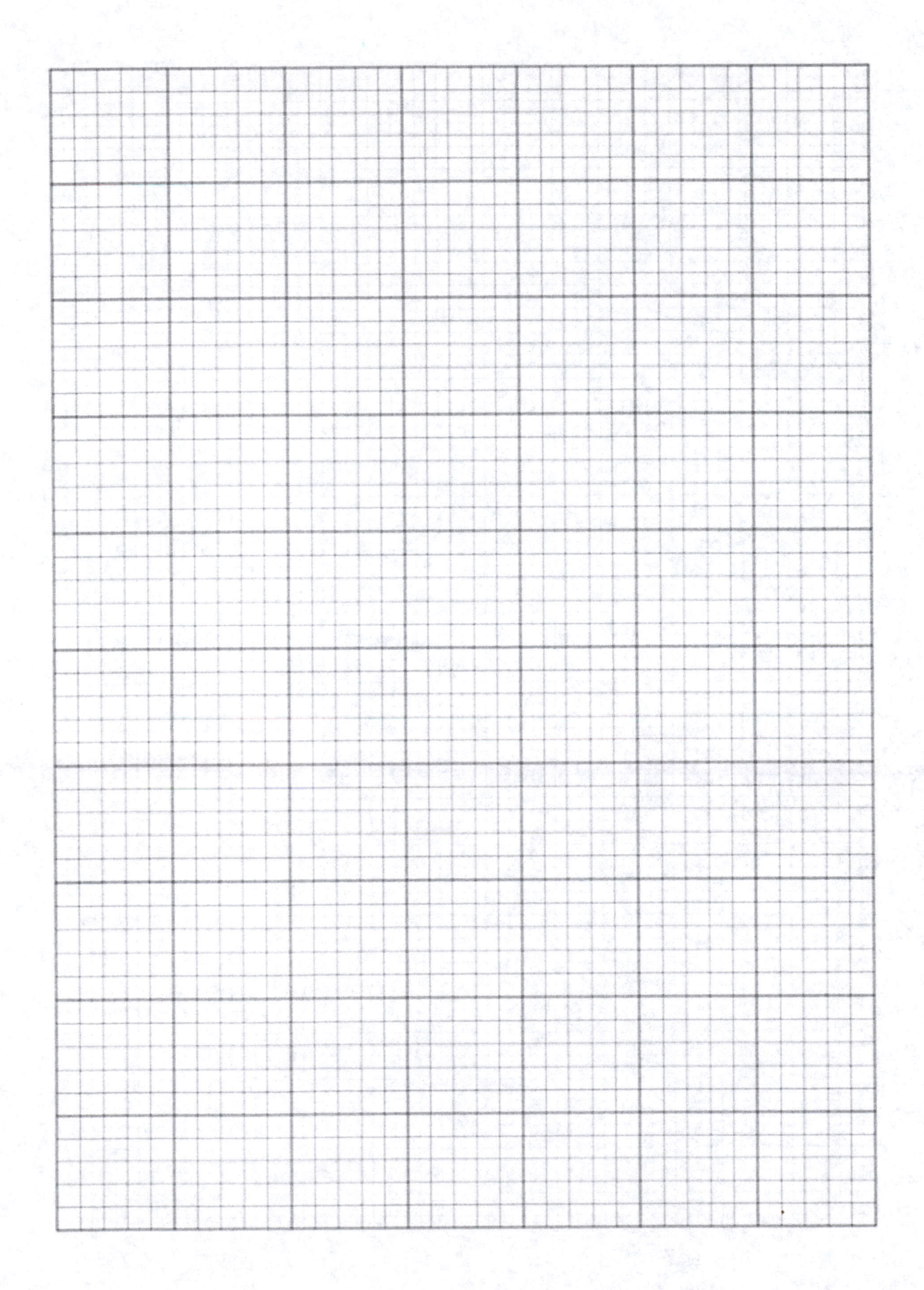

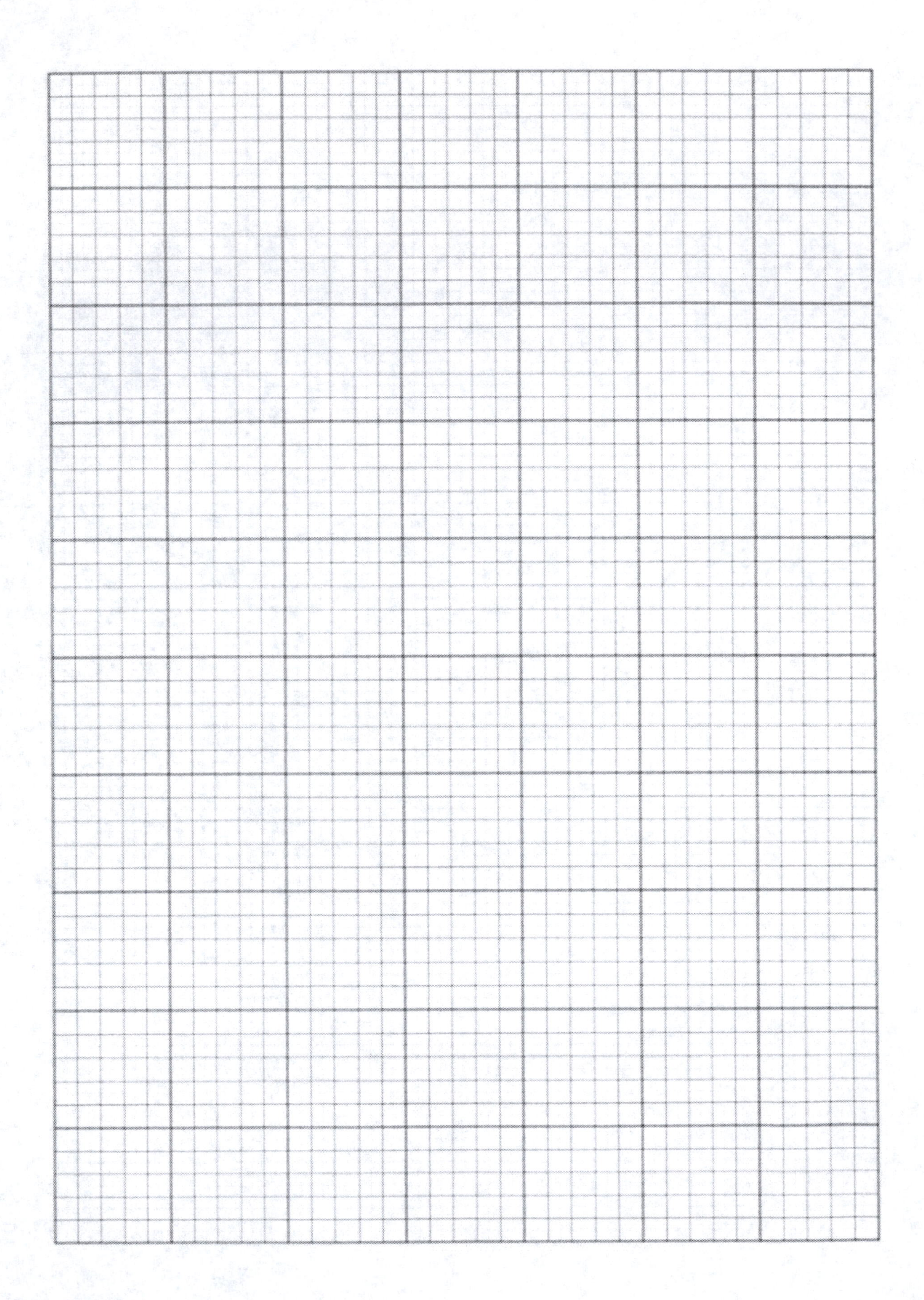

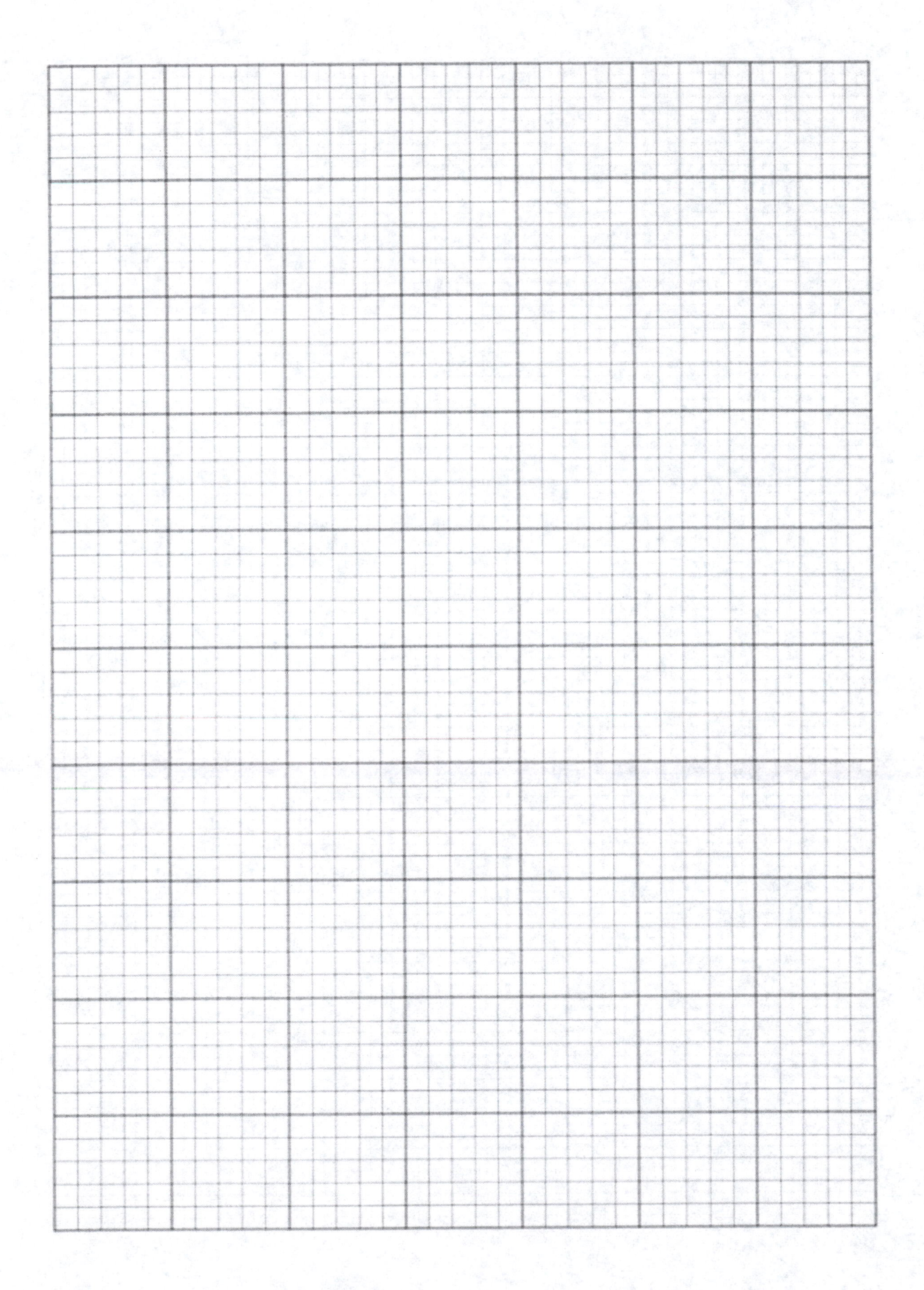

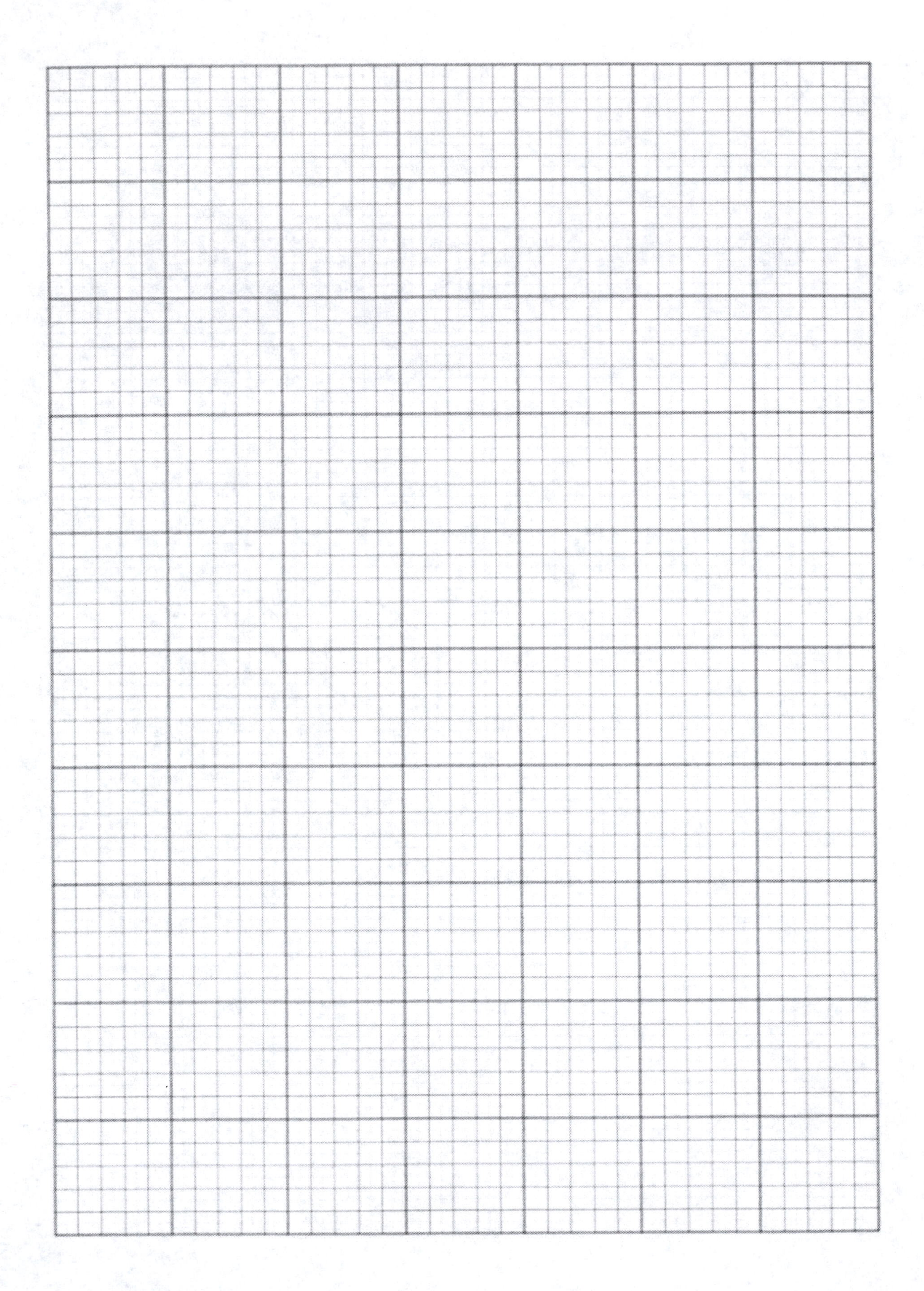

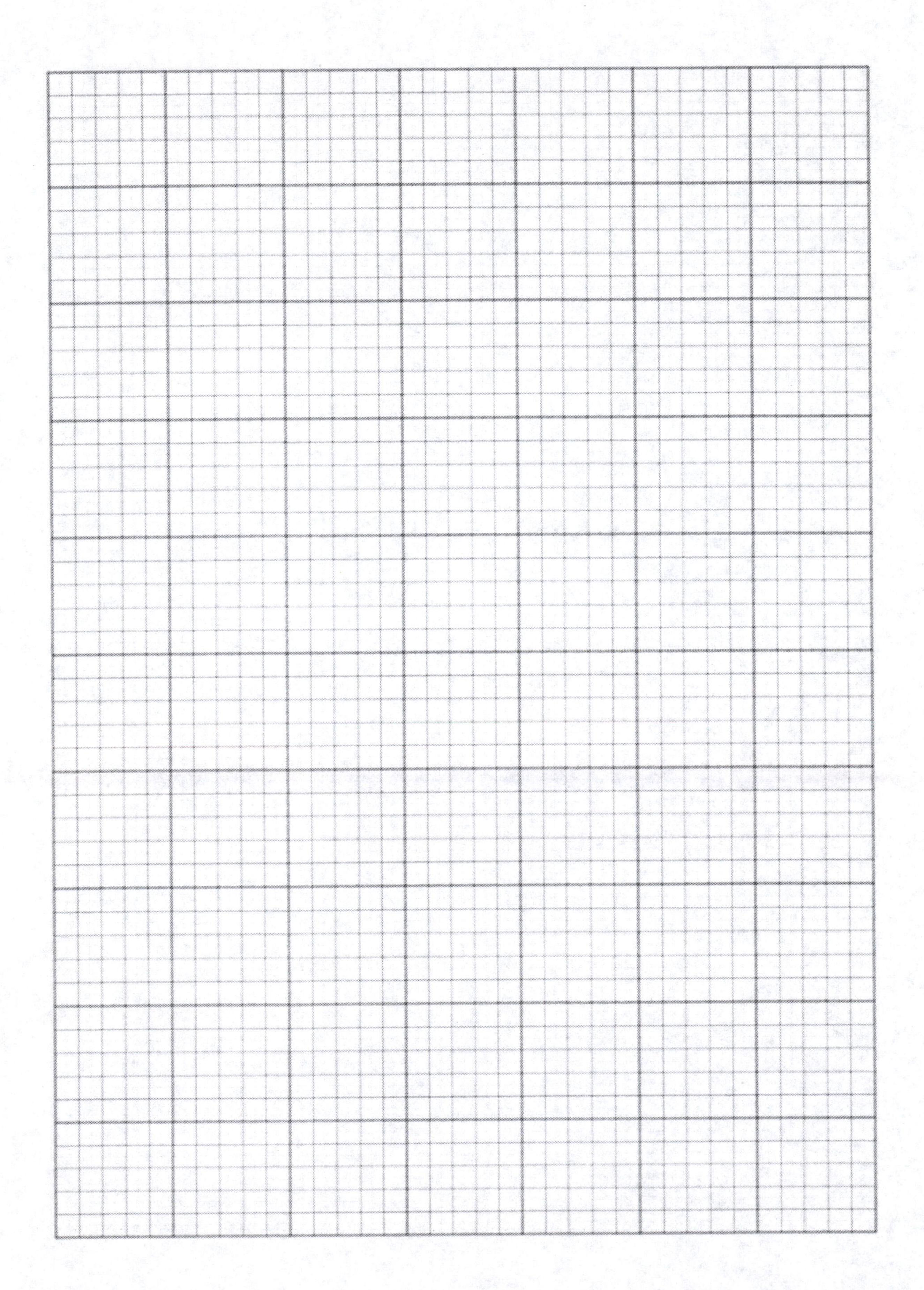

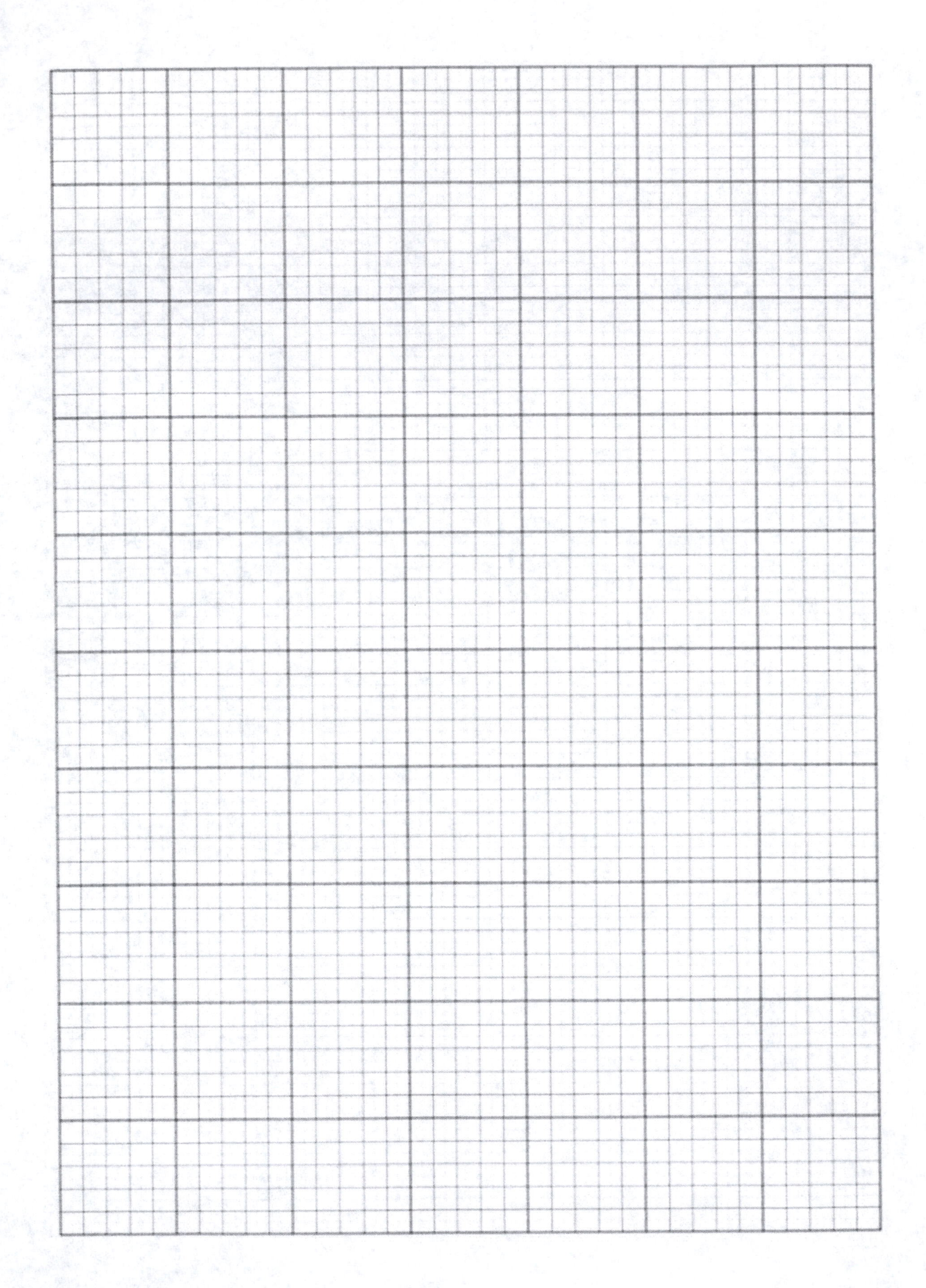

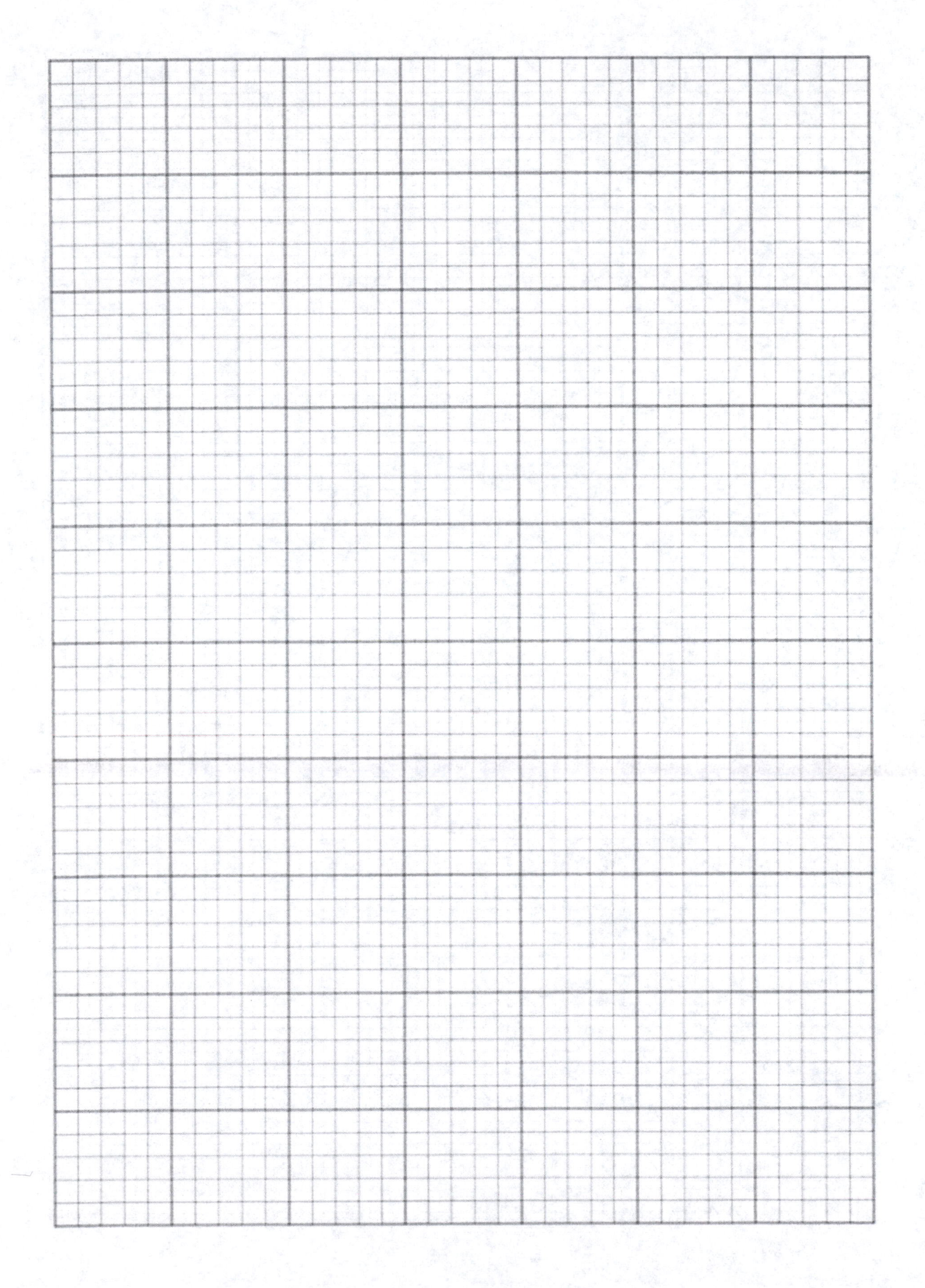

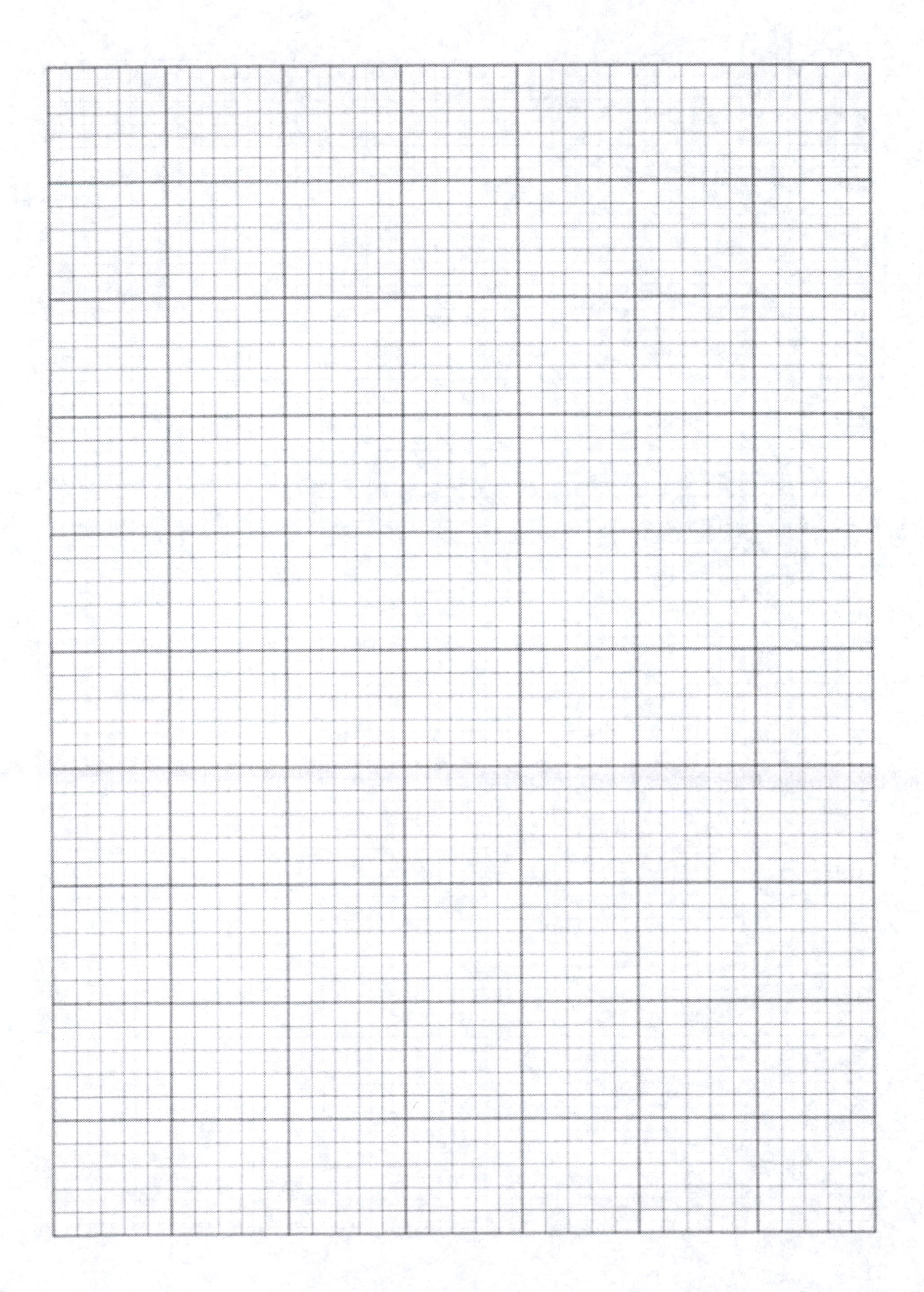

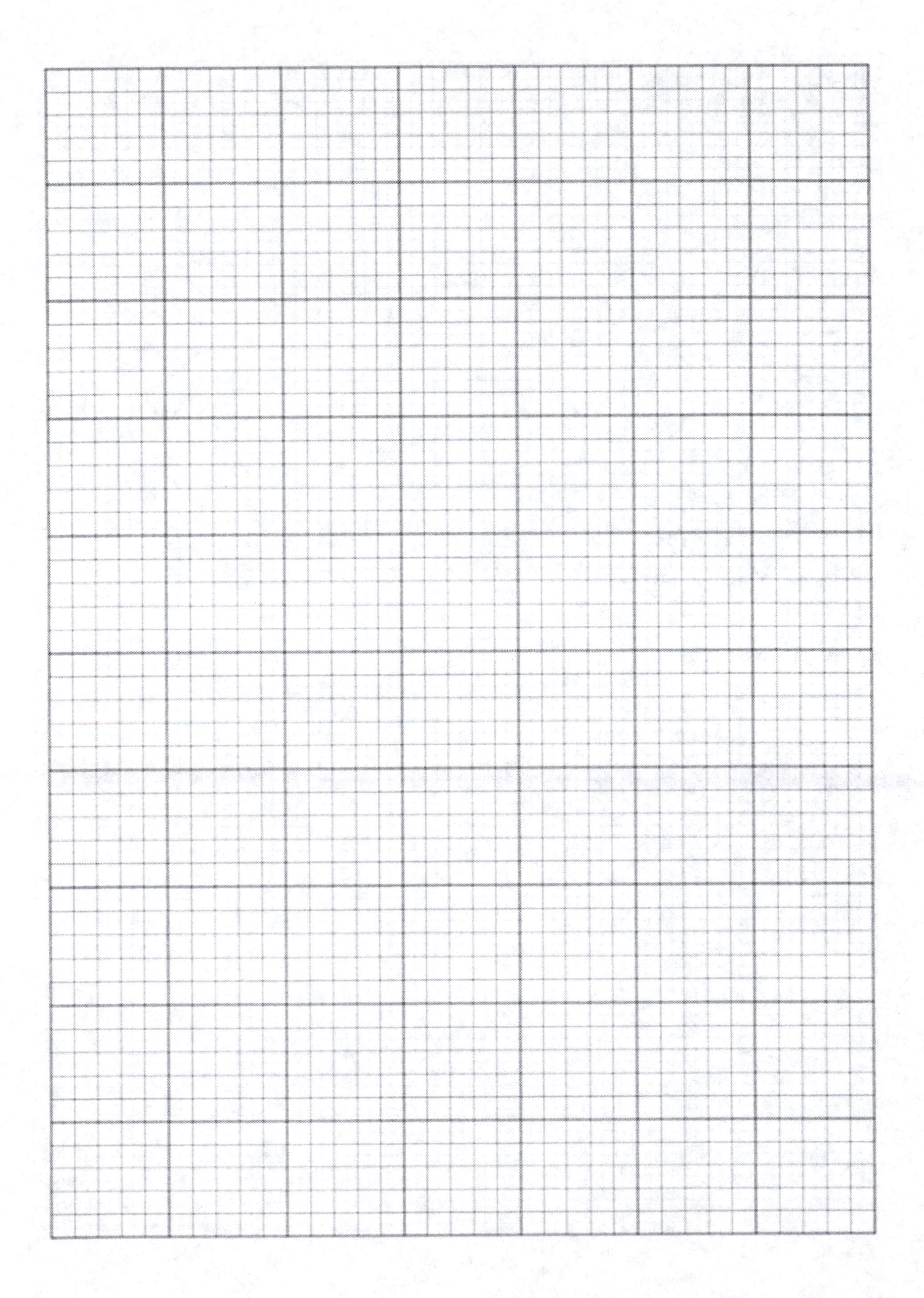

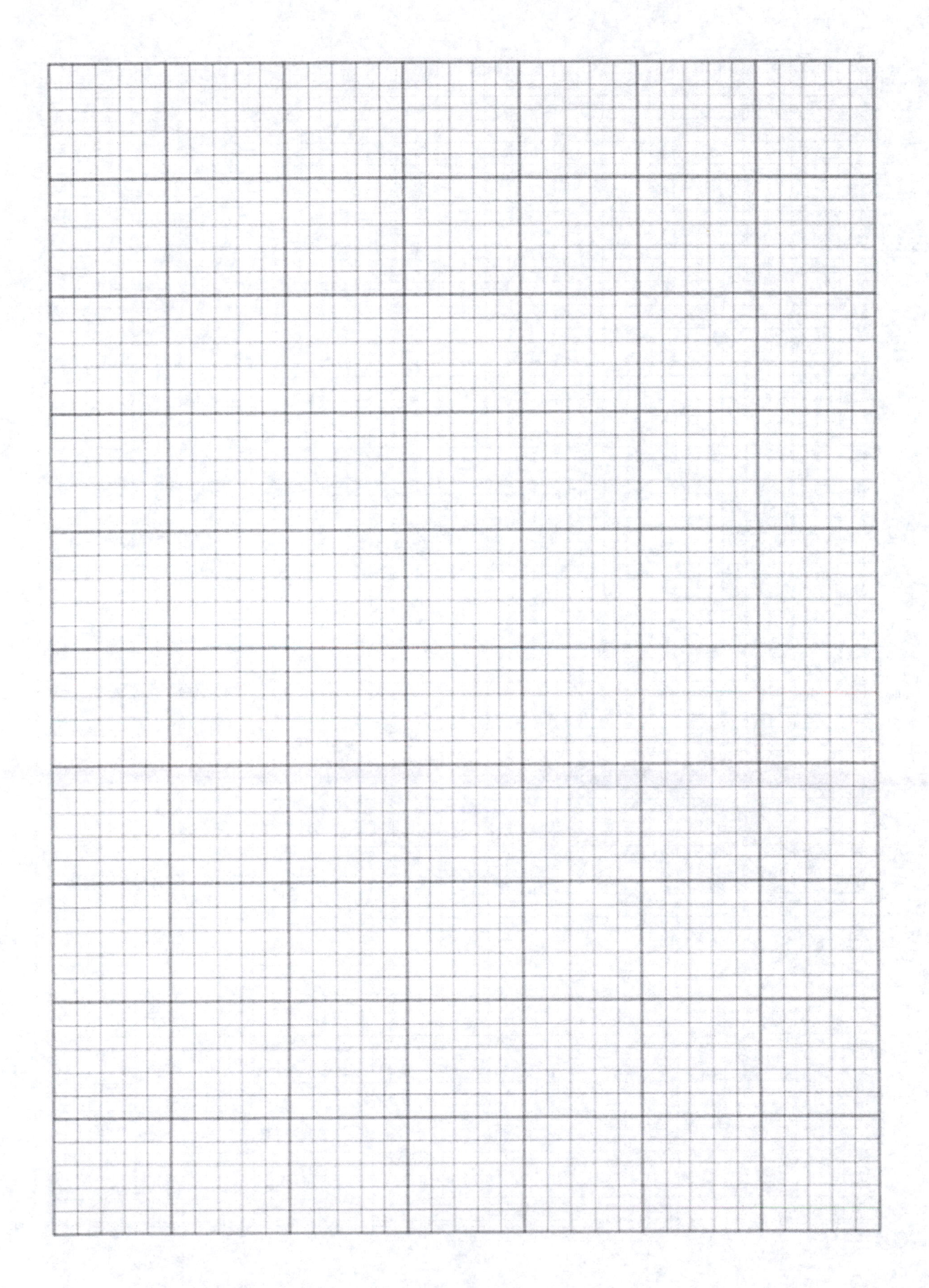

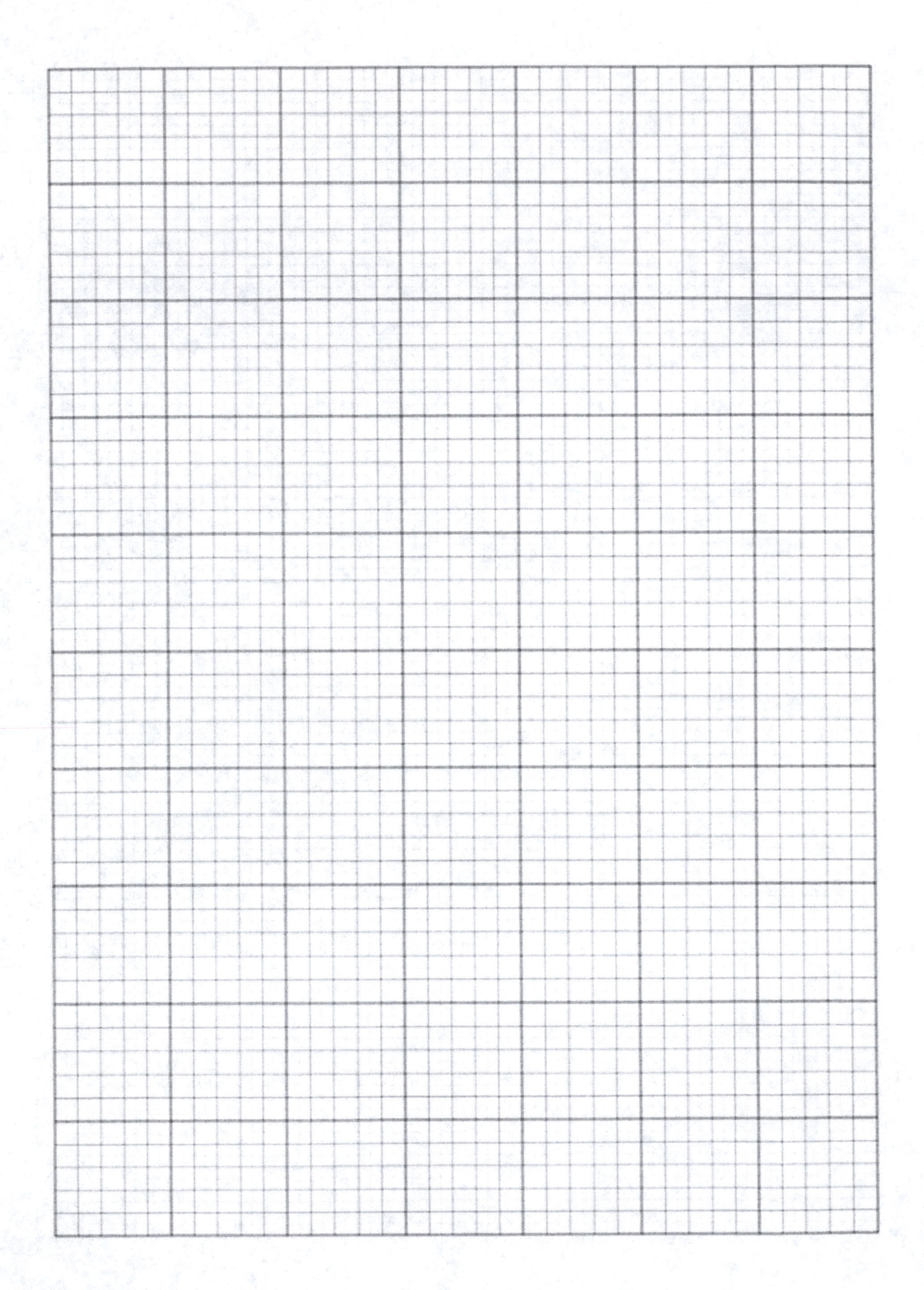

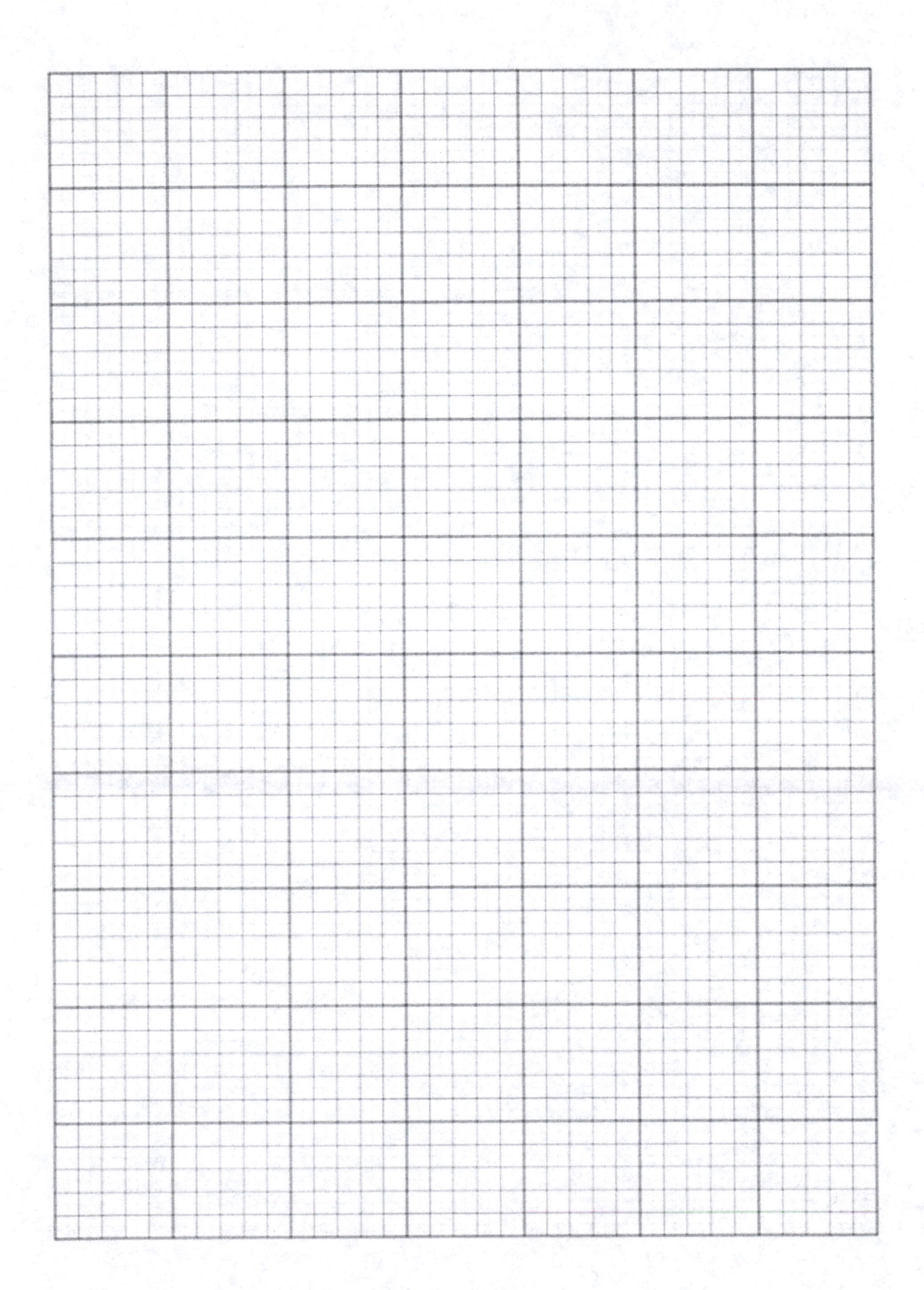

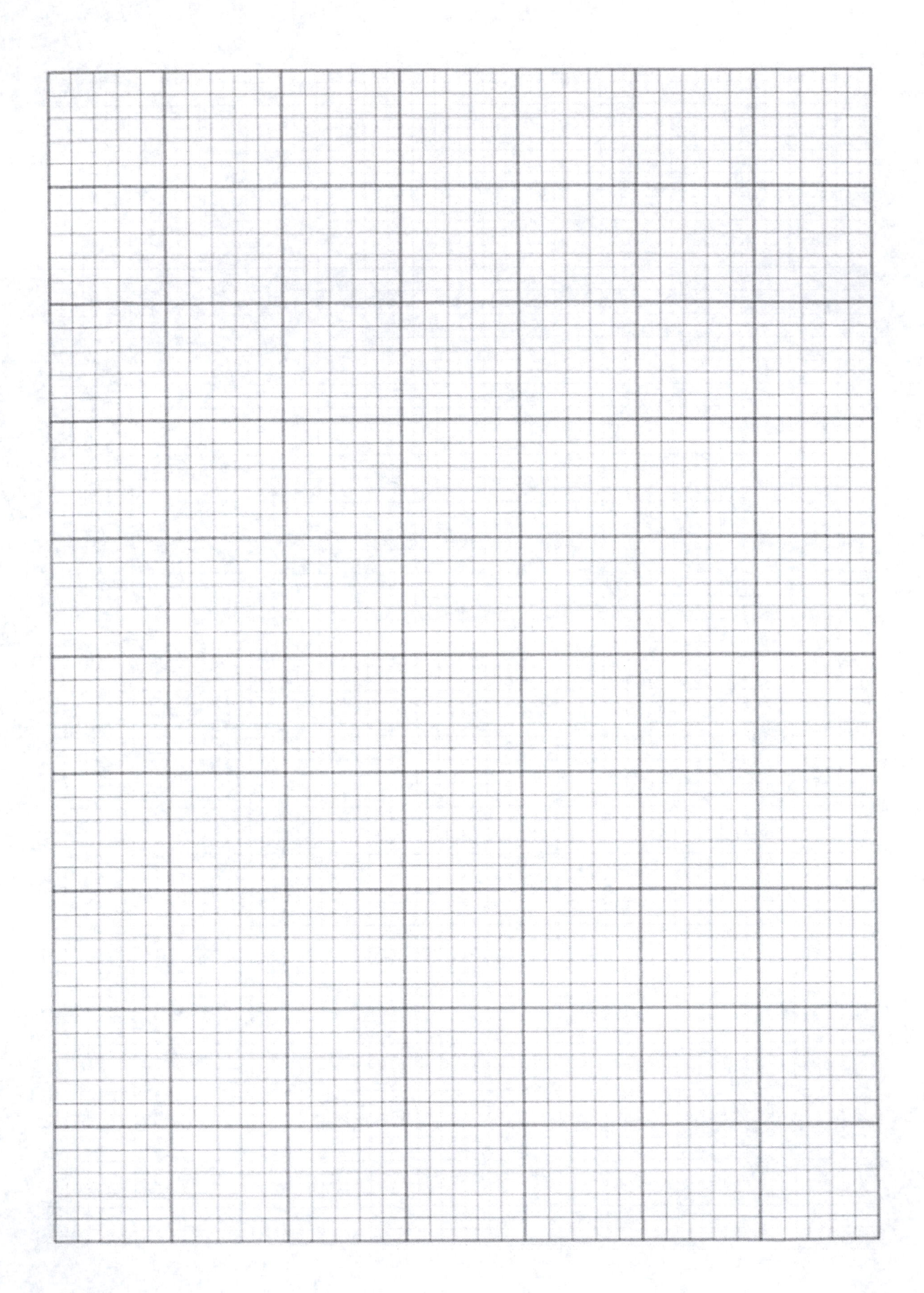

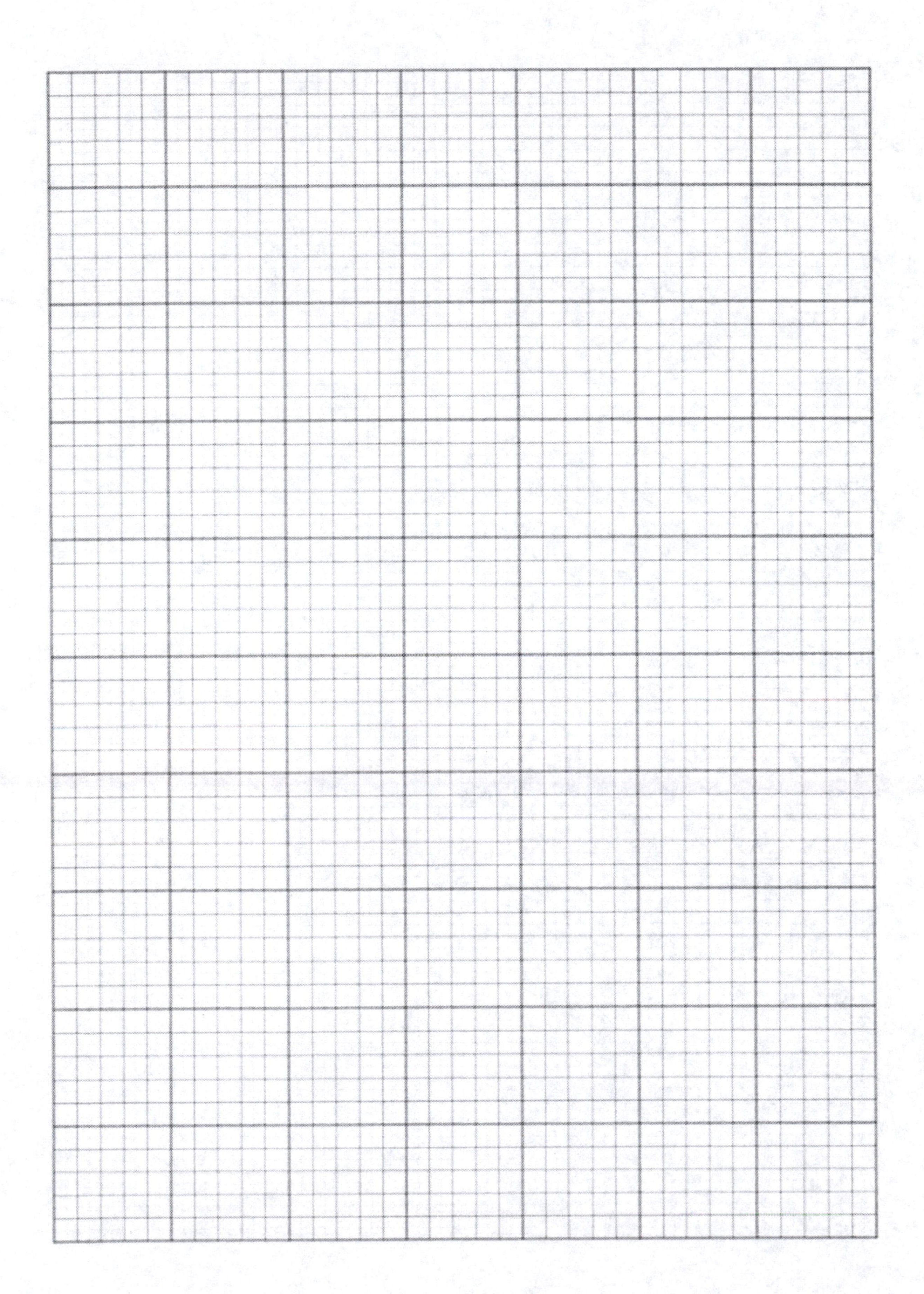

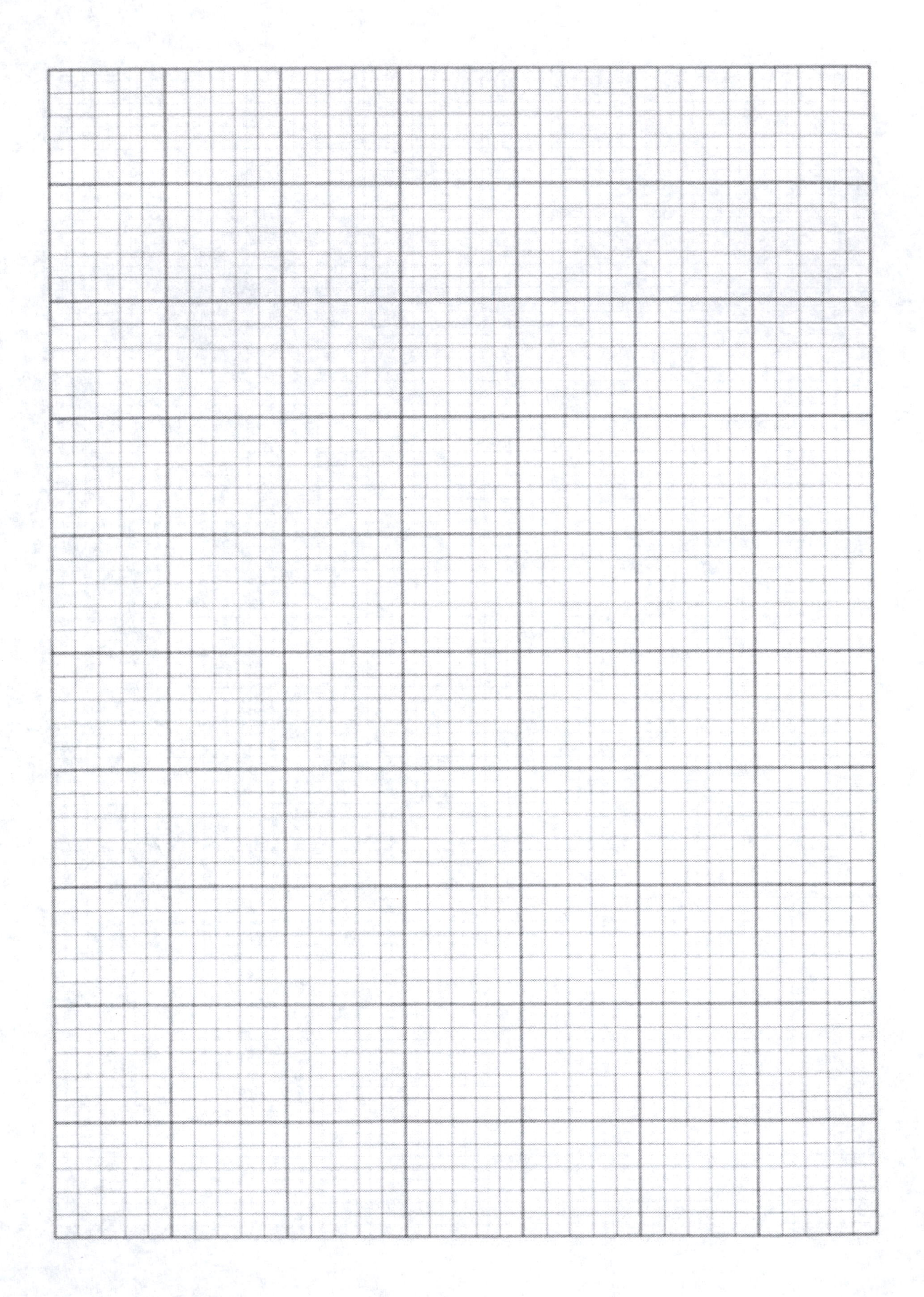

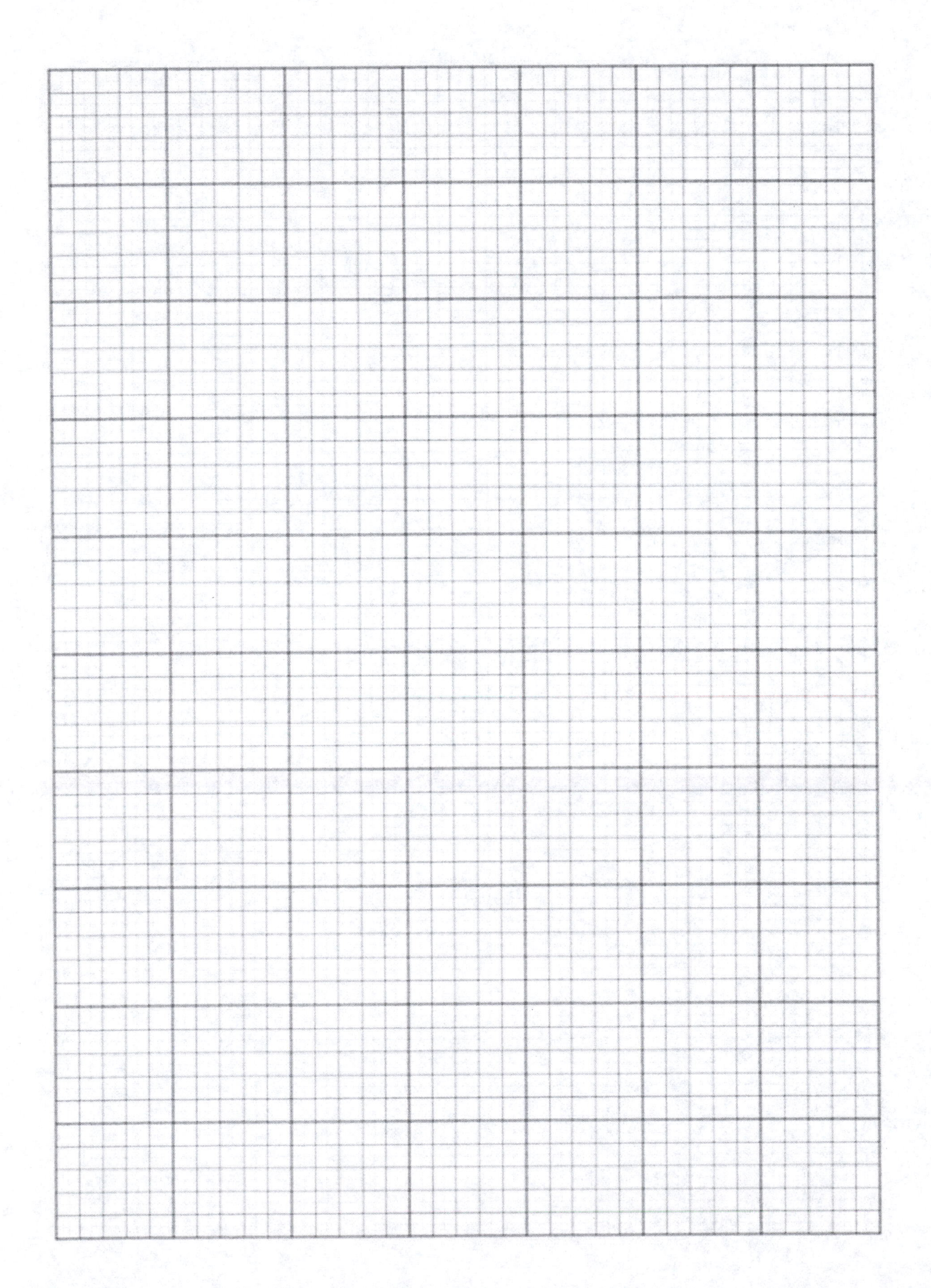